MW01245931

US Critical Infrastructure

Its Importance and Vulnerabilities to Cyber and Unmanned Systems

Dr. Terence M. Dorn

First Edition

PAGE PUBLISHING
Conneaut Lake, PA

First originally published by Page Publishing 2023

ISBN 979-8-88793-062-6 (pbk)
ISBN 979-8-88793-076-3 (hc)
ISBN 979-8-88793-063-3 (digital)

Printed in the United States of America

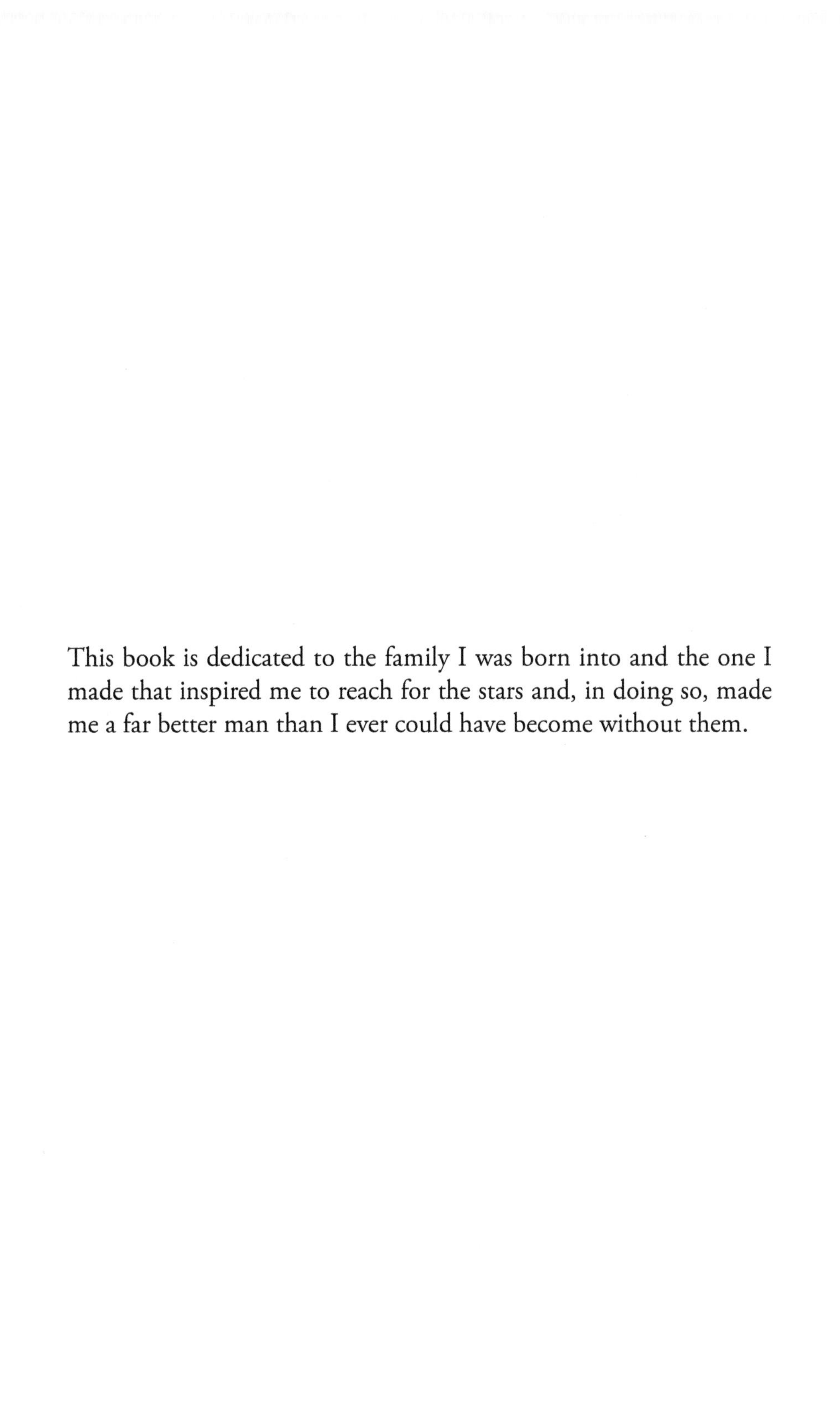

This book is dedicated to the family I was born into and the one I made that inspired me to reach for the stars and, in doing so, made me a far better man than I ever could have become without them.

Contents

Abstract

This book examines the importance of the sixteen sectors of US critical infrastructure and their vulnerabilities to attack by cyber and unmanned systems (aerial, surface [atop water and land], and undersea). No known scholarly study existed that examined the vulnerabilities of one sector of US critical infrastructure to attack by UAS until Phenomenological Study Examining the Vulnerability of U.S. Nuclear Power Plants to Attack by Unmanned Aerial Systems was published late in 2020, and a follow-up book titled *Unmanned Systems: Savior or Threat* was published in late 2021. Technological advancements, such as artificial intelligence and enhanced scientific cyberattacks tools, have combined with unmanned systems to transform the nature of the threat to critical infrastructure. The technological advances with unmanned aerial systems (UAS), unmanned surface systems (USS), and unmanned undersea systems (UUS) have provided additional threat platforms and aids for cyberattacks. Our national security demands strategic leadership and resource allocation by the federal government to rapidly close the critical infrastructure security vulnerabilities that exist.

Chapter 1
Introduction

It has been two years since the publication of my dissertation, the first known scholarly study examining the threat posed by unmanned aerial systems to one sector of critical infrastructure, titled *A Phenomenological Examination of the Vulnerability of U.S. Nuclear Power Plants to Attack by Unmanned Aerial Systems (UAS)*, and a year since *Unmanned Systems: Savior or Threat* was published. I have received many comments from people who have purchased and read one or the other, and they often comment having to do with this material being scary and that they were unaware that such a threat existed. They were also dismayed by the lack of progress on our federal government to keep the public safe and secure our nation's national security. The latter is a continual battle against the innovative actions of our strategic competitors to negate US superiority, reduce US influence while denigrating the individual rights and freedoms that the US affords its citizens as they strive to keep their citizens in check. Much has transpired in the unmanned systems arena since then, and the threat these systems pose to the public and any nation's critical infrastructure has only grown. The growing list of incidents have only served to validate my study and book as the incidents involving the nefarious uses of unmanned systems, namely UAS, have grown worldwide, and the migration of technologies and innovative use by bad actors are only beginning to become commonplace in unmanned surface systems (USS) and unmanned undersea systems (UUS) as well.

Point defense systems designed to counter UAS (C-UAS) approaching specific targets, such as buildings, convoys, public events, have indeed grown. Still, few are applicable for use in a downtown area where the use by law enforcement of jamming an approaching UAS and possibly resulting in it crashing to the ground would pose a severe threat to the public, including the jamming side effects impacting individuals with electronic medical devices inside their bodies, communication systems, business electronics, and to traffic. The widespread defensive schemes such as the Federal Aviation Administration (FAA) establishing no-fly zones and speed and altitude limitations only deter honest, law-abiding people, not criminals and other agenda-driven factions. UAS can be programmed with a fail-safe that will allow them to land softly or return to their start point if it is jammed. If it is not preprogrammed with a fail-safe setting, C-UAS systems could result in it falling uncontrollably to the ground. A small five-pound UAS falling from a height of one hundred to four hundred feet would cause considerable damage on impact to people and property.

In July 2020, a Chinese Da-Jiang Innovations (DJI) manufactured UAS flew toward a "Pennsylvania power substation dangling two 4-foot nylon ropes with thick copper wire connected to the ends of the rope" (Barrett 2021; Lynaas 2021). This attack has only been released by the Department of Homeland Security (DHS) recently, indicating that they had hidden it for a year. Now that a year has passed, law enforcement agencies appear to be stymied in their attempts to discover the culprit. DHS and the Federal Bureau of Investigation (FBI) have pursued this incident as a case of domestic terrorism because the culprit had intentionally removed all identifiable markings, its onboard camera, and memory card from the commercial UAS to avoid identification by the subsequent investigation by law enforcement. The most obvious goal of the culprit was to interrupt the steady supply of electricity to an area by creating a short circuit within that power substation area of influence. Fortunately, in this case, and for reasons unknown to the public, the UAS crashed on the roof of an adjacent building before beginning its attack upon the power substation. The Pennsylvania power station attack represents the first known UAS attack against critical infrastructure targets within the homeland. Based on the previous global incidents, it was only a matter of time before UAS attacks occurred against critical infrastructure. This escalation in the UAS use for nefarious purposes has metastasized and now represents a daily threat to Americans. With the advent of UAS many years ago, it is possible that this was not the first. Indeed, it will certainly not be the last. The federal government withheld information about this incident, just as the Nuclear Regulatory Commission (NRC) did with the UAS overflights of nuclear power plants.

In 2019, a dogged journalist submitted enough Freedom of Information Act (FOIA) requests to force the hands of the NRC. They released a treasure trove of information that showed that in 2020, there were fifty-seven UAS overflights of twenty-six different nuclear power plants dating back five years. This is the face of the new unmanned aerial system threat to the homeland; commercial UAS can be purchased and have identifying serial numbers removed by the operators, which the Pennsylvania culprit did. Even worse, there is an ample supply of parts for the average person to order and build their own UAS. The electronics would presumably not include mandatory software to abide by FAA regulations, and if the parts did arrive with the software installed, it could easily be removed. A long-awaited FAA initiative is geofencing, "a virtual perimeter for a real-world geographic area," that uses a location-awareness when a UAS enters or exits a geofenced area" (FAA n.d.). This crossing of a line action would trigger an alert to the UAS user and the geofence operators. This information could contain the device's location and additional information if the UAS is registered. It is doubtful that bad actors would abide by or care about geofencing, so once again, it will be a means to keep good people honest and will not affect bad actors.

Attacks utilizing UAS have existed for many years and date back to the 2014 use of two UAS to attack a French nuclear power plant in Lyons. One of the two stopped to record the second one as it flew into one of the reactor buildings, where it smashed into multiple pieces and fell to the ground. The building was undamaged.

Sometime during the week of January 10, 2021, UAS were reported flying over the three nuclear power plants in Sweden, and the country's intelligence service assumed control over the investigation. The two nuclear power plants at Forsmark and Oskarshamn are operational, while the third at 151 Ringhals is not (Associated Press, 2021). According to Hans Liwang, an associate professor with the Swedish National Defense College, "Sweden is not sufficiently prepared for this type of event; we have not adapted our way of looking at this type of event to today's reality. We still think of the world as either at peace or war." (Associated Press, 2021).

> "an associate professor with the Swedish National Defense College, he told a Swedish broadcaster that Sweden is not sufficiently prepared for this type of event; we have not adapted our way of looking at this type of event to today's reality. We still think of the world as either at peace or war." (Associated Press 2021)

On January 11, 2022, a Mexican drug cartel operated UAS dropped bomblets on a rival cartel in El Bejuco and La Romera, Mexico (Newdick 2022). Mexican cartels had previously used small UAS (sUAS) to drop single explosives or to perform suicide attacks.

On January 17, 2022, a UAS attack occurred in the UAE at an oil facility in the capital of Abu Dhabi, resulting in three deaths (Qiblawi 2022). Yemen's Iran-backed Houthi rebels claimed credit for the attacks. The UAE and Saudi Arabia air forces responded with airstrikes on the Yemeni capital, killing twelve people (Qiblawi 2022). The UAE's Ministry of the Interior subsequently initiated a ban on UAS and light aircraft as the authorities believe that the bad actors had previously used UAS to conduct reconnaissance of the area and to pinpoint targets for attack.

On November 7, 2021, extremist factions of the Iraqi government utilized three explosive-laden UAS to assassinate Prime Minister Mustafa al-Kadhimi. The assassination attempt appears to be part of an ongoing effort to repudiate the October election results by Iran-backed parties. While the prime minister was unharmed, six security detail members were wounded. It appears that two of the attacking UAS were downed by Iraqi security forces before reaching the presidential residence, and only one reached its intended target before exploding.

In 2015, UAS flew through highly restricted airspace in Washington, DC, and landed on the White House lawn; in 2018, a UAS was used in an assassination attempt against Prime Minister Maduro in Venezuela; in 2019, a disgruntled ex-boyfriend in Pennsylvania used a DJI UAS to drop small explosives and nails on his ex-girlfriend's property as well as her neighbors; in 2019, a well-coordinated attack by cruise missiles acting as motherships launched multiple UAS against Saudi Arabia's Abqaiq refinery, the world's largest oil processing facility, and the nearby Khurais oil field. The Islamic State (ISIS) may have been the first. Other terrorist groups have followed suit in utilizing commercial off-the-shelf consumer quadcopters to conduct surveillance and, after modification, for offensive operations.

The DHS UAS program management office has been primarily focused on answering congressional inquiries and drafting annual reports to Congress. In 2019, I worked in a subordinate directorate, the Countering Weapons of Mass Destruction (CWMD), and led the effort to create the first strategy at DHS for Counter-Unmanned Aircraft Systems (C-UAS) armed with weapons of mass destruction. The strategy was signed by the DHS assistant secretary for CWMD, and within six months, I developed an implementation plan for the approved strategy. The good news about strategies referencing our federal government is that the DOD and DHS are proficient at creating strategies, but the bad news is that they rarely create a strategic implementation plan to accompany the strategies. The implementation plans affix specific responsibility by subordinate division for tasks, timelines for when they need to be completed, and standards by which the tasks can be judged to be completed and to standard. Unlike the signed 2019 strategy for C-UAS armed with weapons of mass destruction, its strategy implementation plan languished in the inbox of a division chief who possessed neither the leadership nor management skills to lead the organization effectively; thus, it was never forwarded to the assistant secretary for signature and never had the impact on DHS that it was designed to do. When I left my position with DHS, I felt that the DHS leadership did not acknowledge the threat posed by unmanned systems to US critical infrastructure, and its citizens had no idea of how to protect the public and were not interested in consolidating resources to counter an immediate threat and was growing. Their concept of searching for viable C-UAS was to spend a small amount of money on research and development (R & D) efforts that were progressing at a snail's pace.

The Pennsylvania attack upon a component of critical infrastructure proves that the UAS threat to the homeland is real. The federal entities responsible for protecting the public have yet to fathom the scale of the danger, not just aerial ones but also unmanned systems that are beginning to operate autonomously on land and atop and beneath our twenty-five thousand miles of waterway. In 2020, an Interagency Security Committee of the US DHS Security Cybersecurity and Infrastructure Security Agency (CISA) produced a document to assist the federal, state, local, tribal, and territorial (FSLTT) echelons of government titled "Protecting Against the Threat of Unmanned Aircraft Systems (UAS)." The document provides basic information on what categories of UAS and recommends minimizing one's presence to UAS. Still, it failed to offer anything substantive on how to counter the threat of UAS and what to do if UAS attacks. Run and hide is simply an abrogation of the federal government's responsibilities by its key responsible agency to aid those echelons of government in desperate need of it who do not have the authority or resources to combat the threat of UAS attacks directly (DHS 2020). In sum, the federal government is doing little to counter the UAS threat to the homeland UAS, USS, and UUS.

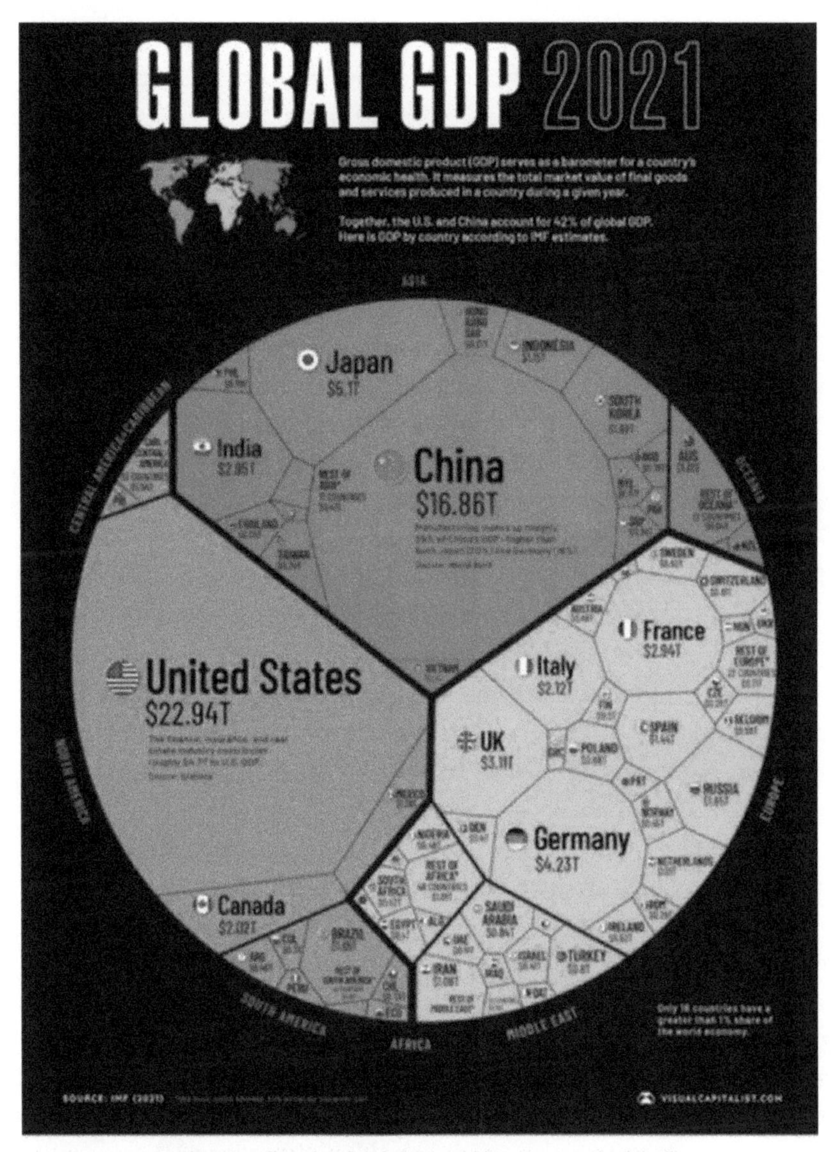

Image courtesy of D. Neufeld at Venture Capitalist, https://www.visualcapitalist.com/visualizing-the-94-trillion-world-economy-in-one-chart/

The newest UAS invention is a laser that has been designed to detonate unexploded ordinance on the ground. The Turkish Eren UAS can fly up to 10,000 feet and recently fired its laser at a range of "328 to 1,640 feet and was able to burn a hole in three millimeters of steel in 90 seconds" (Peck 2021). Rebels in the Central African Republic took a plastic bottle, a piece of string, and a hand grenade and built an aerial platform that has inflicted significant losses on their enemies (Hambling 2021). This is indicative of early twenty-second-century warfare, sending one's unmanned systems to kill the enemy, in this case with high-tech UAS. ISIS was the first group in 2016 to take a DJI UAS, add a simple release mechanism and a hand grenade, and turn it into a cheap aerial platform. Multiple organizations in the Middle East are now arming UAS and using them against their enemies, including the Iraqi Federal Police and drug cartels in Mexico. They have used them against rival gangs and the police (Hambling 2021). Russia has been testing its Orion UAS in an offensive role to engage rotary-wing targets. Thus, it can also function as a UAS killer (Newdick 2021). So this is the predicament the US now finds itself in. The most significant economic and military powerhouse on the planet is in the untenable position of having over twenty thousand critical infrastructure facilities vulnerable to attack by UAS or cyber.

Two allied nations used two C-130 cargo aircraft to launch autonomous Silent Arrow GD-2000 gliders in the Middle East. Each navigated itself to a landing point with 1,026 pounds of cargo for 50 percent less than manned cargo aircraft (Tingley 2022). There is simultaneous work being performed on "helicopter" gliders that have a range of forty miles once released from manned aircraft. These could easily provide humanitarian support to areas affected by disasters and military support to units on the move to include naval forces that are underway at sea. It is conceivable that such aerial craft could also be used for nefarious purposes such as delivering explosives and munitions to bad actors that have infiltrated a nation and are planning attacks, either against military, political, or critical infrastructure facilities.

The following sixteen chapters will describe the importance of each sector of US critical infrastructure, its interdependence with other sectors, and its vulnerabilities. I will examine it utilizing risk, which has been defined in the National Infrastructure Protection Plan (NIPP) as "a function of consequences, human and economic, vulnerabilities, and threats" (CISA 2013). According to Rick Moy:

> "domestic critical infrastructure is arguably now more at risk than at any point in living memory, and certainly in a peacetime context. The highly digitized and connected nature of today's healthcare, public sector, building, and industrial systems have placed them firmly in the sights of cybercriminals and malign foreign powers. There can be no doubt that protecting critical infrastructure is vital, given that 70% of breaches in 2019 were perpetrated by external actors, according to the Verizon 2020 DBI Report. The majority (64%) are financially motivated, with espionage accounting for 5% of breaches. The cybersecurity challenges facing critical infrastructure vary by sector, and part of the problem is that there is no 'catch-all' remedy. In the energy and water utility sectors, for example, their IoT and OT networks are critical to both the delivery of vital public services and to cash flow. If critical infrastructure experiences an outage, there is a measurable negative impact on the organization and customers alike. In the financial services sector, institutions have become high-value targets of cyber-attacks and must deal with a range of issues, including securing remote users, managing network access, and addressing compliance requirements from regulators. Healthcare organizations, by contrast, are focused on patient privacy and data access on open networks, security

and privacy for medical devices and systems, and the security of micro-segmentation for healthcare workers, vendors, contractors, and third parties. If that wasn't enough, during the COVID-19 period, cybercrime has quadrupled, according to the FBI, while the World Health Organization, for example, reported a fivefold increase "in the number of cyberattacks directed at its staff and email scams targeting the public at large. As a consequence of the pandemic, there have also been multiple attacks on electricity grids, water systems and energy organizations, election locations, and newly distributed enterprises. To an extent, the rise in cybercrime in these areas is a consequence of the rapid pace of IT innovation, and the widening network perimeters that require protection. The development of critical and physical infrastructure has created attack vectors that didn't exist when many of today's most popular security solutions were being designed. The result is rapidly rising cost and complexity, with a measurable decline in protection. The original inventors of the networking technologies that became the internet never imagined the billions of devices we use today. They introduced something that turned out to be a huge, critical assumption: IP addresses could safely be used as both location and identity. Every device gets an IP address, that address is used as its identity and can be found. If it can be found, it can be hacked. Moreover, every device hacked can be used to hack other devices. The sudden and widespread shift to remote working environments, for example, has laid bare the limitations of some dominant security solutions. Critical infrastructure and distributed organizations sit on the front line of the cybersecurity ecosystem. Protecting them must represent a line in the sand in the fight against malicious activity—failure to do so will have widespread and long-term consequences for large sections of society." (Moy 2020)

Chapter 2
Overview of US Critical Infrastructure

DHS defines critical infrastructure as a physical or virtual asset, system, or network whose "destruction or incapacitation would harm national security, public health, safety, economy, or any combination of them: it must be operational for citizens to thrive and be successful (DHS, n.d.). On February 12, 2013, President Obama signed the Presidential Policy Directive 21 (PPD-21), which identified the following sixteen sectors of critical infrastructure and affixed responsibility to sector-specific agencies (SSA) (The White House 2013):

1. Chemical: DHS
2. Commercial facilities: DHS
3. Communications: DHS
4. Critical manufacturing: DHS
5. Dams: DHS
6. Defense industrial base: Department of Defense (DOD)
7. Emergency services: DHS
8. Energy: Department of Energy (DOE)
9. Financial services: Department of the Treasury (USDT)
10. Food and agriculture: US Department of Agriculture (USDA) and the Department of Health and Human Services (HHS)

11. Government facilities: DHS and the General Services Administration (GSA)
12. Healthcare and public health: HHS
13. Information technology: DHS
14. Nuclear reactors, materials, and waste: DHS
15. Transportation systems: DHS and the Department of Transportation (DOT)
16. Water and wastewater systems: Environmental Protection Agency (EPA)

A numerical breakdown of the responsible federal government agencies in the entirety of the sector-specific critical infrastructure or collaboration with other federal entities is as follows: DHS-10; DOD-1; DOE-1; DOT-1; USDA-1; EPA-1; GSA-1; and HHS-2. Clearly, the bulk of the homeland's security responsibility resides with the DHS.

According to the White House:

> "the nation's critical infrastructure provides the essential services that underpin our society. Proactive and coordinated efforts are necessary to strengthen and maintain secure, functioning, and resilient critical infrastructure, including assets, networks, and systems vital to public confidence and the nation's safety, prosperity, and well-being. The nation's critical infrastructure is diverse and complex. It includes distributed networks, varied organizational structures and operating models (including multinational ownership), interdependent functions and systems in physical space and cyberspace, and governance constructs involving multi-level authorities, responsibilities, and regulations." (The White House 2013)

Critical infrastructure owners and operators are precluded by their federal agencies or departments to procure the necessary equipment to sense UAS threats flying toward their facilities and countering them with jammers or other C-UAS systems. In an about face, the NRC released a treasure trove of UAS-related incursions that showed that in the past five years, there had been fifty-seven UAS incursions over twenty-six separate nuclear power plants. The facilities' security force was left to stare up in the air, watch the pretty lights as the UAS flew over and in between buildings, and to report the incident to their headquarters. The reason? Simple: after a three-year security review, the NRC determined that UAS pose no significant threat to nuclear power plants, and thus, the owner and operators are precluded from purchasing sensors to detect inbound UAS and C-UAS such as jammers, to deter or bring them down to the ground. No federal department or agency has publicly acknowledged the threat posed by USS nor UUS to the homeland and its waterways and roads and critical infrastructure facilities. According to the White House and DHS, critical infrastructure must be secure and withstand and rapidly recover from all hazards" (The White House 2013; DHS, n.d.), yet with thousands of facilities and locations across the US, there is simply no way that they can all be secured.

A great first step would be to authorize the owners and operators of these facilities to defend themselves. According to the White House:

> "achieving this will require integrating the national preparedness system across prevention, protection, mitigation, response, and recovery. The endeavor is a shared responsibility among the FSLTT entities and public and private owners and operators of critical infrastructure. The federal government also has a responsibility to strengthen the security and resilience of its critical infrastructure, for the continuity of essential national functions, and to organize itself to partner effectively with and add value to the security and resilience efforts of critical infrastructure owners and operators." (The White House 2013)

It is the stated policy of the US federal government to,

> "Strengthen the security and resilience of its critical infrastructure against both physical and cyber threats. The federal government shall work with critical infrastructure owners and operators and FSLTT entities to take proactive steps to manage risk and strengthen the security and resilience of the nation's critical infrastructure, considering all hazards that could have a debilitating impact on national security, economic stability, public health and safety, or any combination thereof. These efforts shall seek to reduce vulnerabilities, minimize consequences, identify, and disrupt threats, and hasten response and recovery efforts related to critical infrastructure. The federal government shall also engage with international partners to strengthen the security and resilience of critical domestic infrastructure and critical infrastructure located outside of the U.S. depends on the nation. U.S. efforts shall address the security and resilience of critical infrastructure in an integrated, holistic manner to reflect this infrastructure's interconnectedness and interdependencies. This directive also identifies energy and communications systems as uniquely critical due to the enabling functions across all critical infrastructure sectors." (The White House, 2013)

The three strategic imperatives that the federal governments deem necessary to strengthen national security and build resilience within our sixteen sectors of critical infrastructure is to "Refine and clarify functional relationships across the federal government to advance the national unity of effort to strengthen critical infrastructure security and resilience; [to] enable effective information exchange by identifying baseline data and systems requirements for the federal government; [and finally to] implement an integration and analysis function to inform planning and operations decisions regarding critical infrastructure." (The White House 2013)

The Secretary of the Department of Homeland Security is the cabinet level individual responsible for building resilience within our critical infrastructure. The national effort must include getting the right people with expertise, not political appointees who contributed money or time to the election campaign of the presidential victor; it must include subject matter experts outside of government; day-to-day engagement from all the SSAs and specialized or support capabilities from other federal departments and agencies; and finally, building a true team effort with the private sector critical infrastructure owners and operators and FSLTT entities. This is the strategy that is required to strengthen the security and resilience of the nation's critical infrastructure.

Chapter 3
Chemical Sector

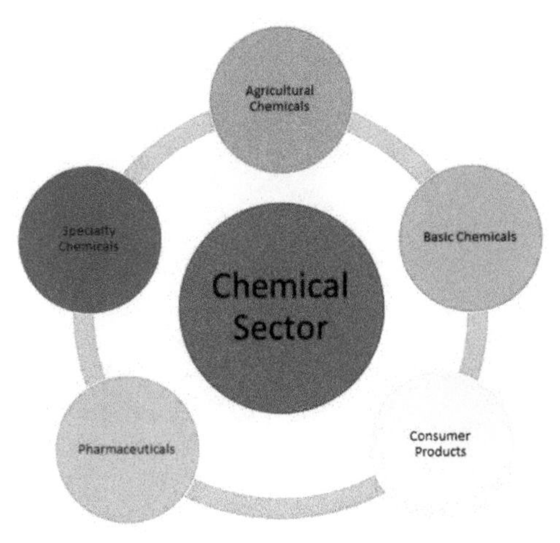

Figure 1: Chemical Sector Major Components

According to 2019 statistics furnished by CISA, the critical infrastructure "chemical sector converts raw materials into more than 70,000 diverse products essential to modern life and distributes those products to more than 750,000 end-users throughout the nation" (CISA 2019a). Furthermore, thousands of chemical facilities within the US use, manufacture, store, transport, or deliver chemicals globally (CISA 2019a). The products produced by the chemical sector and transported throughout the nation are essential throughout the other fifteen sectors of US critical infrastructure, elevating this sector to one that is indicative of its importance to the nation's economy and national security. The US chemical industry produces "15% of the global production; it is a $768 billion industry that accounts for 25% of the U.S. Gross Domestic Product (GDP); and in 2016, 96% of all U.S. goods manufactured used chemical sector products during the manufacturing process" (CISA 2019b). The chemical sector employs "800,000 people; the value of chemical sector exports amount to $174 billion, or ten cents of every export dollar" (CISA 2019b). The chemical sector directly feeds these "Functional areas: manufacturing plants which convert raw materials into intermediate and end products; warehouse storage which provides downsized repackaging and bulk storage; transportation system which transports chemicals to and from manufacturing, plants, warehouses, and end-users." (CISA 2019b)

Figure 2: U.S. Chemical Manufacturing Facilities (2016)

Image courtesy of CISA, www.cisa.gov/sites/default/files/publications/Chemical-Sector-Profile_Final%20508.pdf

The majority of US critical infrastructure facilities, including the chemical sector ones, are privately owned, and operated, thus national security and increasing resilience is dependent upon the FSLTT echelons of government and the private sector to collaborate and execute effective sector strategies. In 2010, chemical shipments amounted to "$720 billion; the chemical industries provided over $226 billion to the U.S. GDP, or 2% of the entire GDP" (Bureau of Economic Analysis, n.d.). In 2010, the chemical sector exported

"$171 billion in chemicals, which was 10% of all U.S. exports, which made it the second-largest U.S. manufacturing sector" (Bureau of Economic Analysis, n.d.). Production of basic chemicals is geographically centralized "along the Gulf Coast, where petroleum and natural gas and refineries are vast; petrochemicals are produced predominantly in Texas and Louisiana (Bureau of Economic Analysis, n.d.). The production of plastics, pharmaceuticals, and fertilizers are dispersed throughout the US.

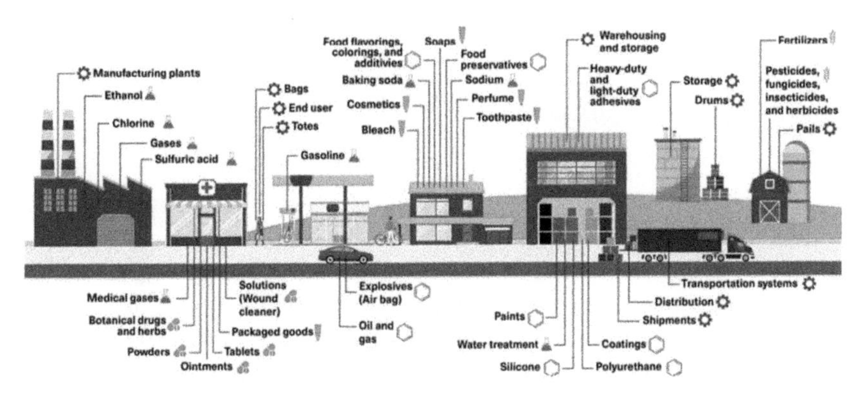

Image courtesy of CISA, www.cisa.gov/sites/default/files/publications/Chemical-Sector-Profile_Final%20508.pdf

The US chemicals industry's "end-use energy consumption, excluding electricity generation, transmission, and distribution losses," totaled 5.2 quadrillion of British thermal units (BTU) "in 2006, accounting for about 24% of energy use across the entire spectrum of U.S. manufacturing" (Energy Information Administration 2006). Since 1974, the chemical sector "has reduced energy consumption by more than half" (American Chemistry Council 2011c). The chemical sector is one of the cornerstones of the US economy; it converts "raw materials, oil, natural gas, air, water, metals, and minerals, into more than 70,000 assorted products (American Chemistry Council 2011b). A great many chemicals are required to create various consumer goods and "thousands of products that are essential to agriculture, manufacturing, computing, telecommunications, construction, and service industries (American Chemistry Council

2011b). In 2010, the chemicals sector "spent approximately $55 billion on R&D; the majority of chemical R&D dollars is spent on product development; basic and specialty chemical companies typically allocate 1% to 3% of their annual sales toward R&D" (American Chemistry Council 2009; American Chemistry Council 2011b). In the US, the chemical sector "employ 784,000 people and indirectly contribute 4.4 million jobs to the economy; the average wage for an employee in the chemistry industry was $81,900" (American Chemistry Council 2011a).

The chemical sector consists of five major components: agricultural chemicals, basic chemicals, specialty chemicals, consumer products, and pharmaceuticals (CISA 2019b). Each component sustains specific components of the nation's chemical requirements.

The agricultural component

The agricultural chemical industry was created to supply chemicals that the farming industry would require, such as fertilizers, pesticides, herbicides, and other chemicals. In 2019, it had "$29.6 billion in sales to the chemical industry; a total of 471 facilities; $41.1 billion in the value of shipments distributed; and 35,100 employees" (CISA 2019b). In 2012, the US farming industry spent "$9 billion on pesticides; $1.4 billion on fungicides; $2.2 billion on insecticides; $.14 billion on fumigants; and $5.1 billion on herbicides" (CISA 2019b). In 2014, the US used 23.2 million tons of fertilizer for farming and home use (CISA 2019b).

Basic chemical component

The basic chemical component produces organic chemicals that create "dyes, plastics, petrochemical products, and inorganic chemicals for solid and liquid industrial gases, sulfuric acid, and chlorine" (CISA 2019a). To do this, "1,277 facilities with 151,700 employees produce $271 billion in product value, including petrochemicals, plastic resins, manufactured fibers, and synthetic rubber" (CISA 2019a). "Two hundred thirty plants producing petrochemicals in the U.S. needed to manufacture car parts, medical devices, water bottles, and food packaging containers" (CISA 2019a). Over "41 million tons of sulfuric acid are produced annually, 16% of the world's production, used to manufacture fertilizers, metal processing, phosphates, fibers, paints, pigments, pulp, and paper" (CISA 2019a).

"Over twelve million tons of chlorine are produced annually, used in the manufacturing of PVC, solvents, water sanitation, and the pulp and paper industry" (CISA 2019). Over "12.2 billion worth of products are produced annually in applications such as medical, industrial, electronics, and food and beverage industries" (CISA 2019). There are "60,000 employees in this sector, and each job generates 2.1 jobs elsewhere in the economy, contributing 24.3 million to the U.S. economy" (CISA 2019).

Specialty chemicals component

Specialty chemicals are manufactured for unique performance or function, including adhesives and sealants, food additives, flavors and fragrances, and explosives (CISA 2019b). The adhesives subcomponent includes resins, glue, rubber, polyurethane, natural rubber, paints, silicone, starch, and others (CISA 2019b). The Food and Drug Administration (FDA) lists 3,000 food additives (CISA 2019b). The flavors and fragrances created $1.4 billion in sales in 2002. Finally, in 2015, the US used "4.5 billion tons of explosives for mining, oil and gas extraction, and vehicle airbags" (CISA 2019b). The total value of "specialty chemical shipments amounted to $89.6 billion" (CISA 2019b).

Consumer product component

In the US, consumer products consist of household products, including "soaps, detergents, bleaches, toothpaste, cosmetics, perfume, and paints" (CISA 2019b). The consumer product component has 2,268 facilities and 108,200 employees. This component surpasses $446 billion in sales, making the US consumer goods market the largest globally, with $74.4 billion in the value of products shipped (CISA 2019b).

Pharmaceutical component

The pharmaceutical component includes "manufacturing, extraction, processing, purification, and the packaging chemical materials as medications" (CISA 2019a). This sector consists of "2,366 facilities, 247,270 employees and generates a total value of shipments of $202 billion" (CISA 2019a). The sector has four subsectors: "pharmaceutical preparation, biological products; diagnostic substances; and medicines" (CISA 2019a). Pharmaceutical preparation consists of "1,315 facilities, 147,510 employees, and $152 billion in the total value of shipments distributed" (CISA 2019a). Biological products consist of "301 facilities, 43,410 employees, and $26 billion in the total value of shipments distributed" (CISA 2019b). Diagnostic shipments consist of "234 facilities, 27,490 employees, and $12 billion in the total value of shipments distributed" (CISA 2019b). Lastly, medicines consist of "516 facilities, 28,860 employees, and $12 billion in the total value of shipments distributed" (CISA 2019b). The top five states' pharmaceutical facilities are California with 430, New Jersey with 161, New York with 142, Florida with 135, and Texas with 116 (CISA 2019b).

The US federal government identified several departments and agencies responsible for monitoring the chemical sector. CISA, a component of DHS:

> "regulates 3,355 facilities; the Transportation Security Administration (TSA) has inspection apparatus set up at 46 key urban areas for rail security; the USCG manages 3,200 facilities; the Environmental Protection Agency (EPA) regulates 800,000 facilities; the Department of Transportation (DOT) requirements impact 13,829 shippers; the Department of Health and Human Services (HHS) regulates $1 trillion in

> consumer products; the Department of Labor conducts 700 inspections of chemical manufacturing facilities; and the Department of Justice, Bureau of Tobacco, Firearms and Alcohol, regulates and licenses 9,815 individuals and businesses." (CISA 2019b)

Threats to the chemical sector

There are natural and man-made threats to this sector. The natural hazards were highlighted in 2017 when the US was hit with sixteen separate billion-dollar disasters, with the total damages estimated to be $300 billion (CISA 2019a). The consequences of natural disasters can be long-term physical ones, with recovery and rebuilding taking months or years. Disasters can also create a disruption of access, power, and fuel loss to chemical facilities; a cascading effect would be supply chain disruptions, which includes the arrival of raw material to the chemical facilities that initiates a variety of industrial processes. Other natural disasters can include earthquakes, flooding, transportation disruptions, tropical cyclones, and hurricanes.

Man-made threats stem from cyberattacks, but these are often long-term campaigns by nation-states. Malware and ransomware have been a favorite of these attacks, and in 2017, Symantec announced that it had discovered 670 million new versions of malware (Symantec 2018). Attacks on cloud services, such as hyperjacking, attacking the software that manages "Virtual machines, escalation, cyber threat breaks out of the virtual environment to gain elevated access, and virtual machine escape, the cyber threat escapes a virtual machine and interacts directly with the virtual machine's hosting environment, [have been seen and were believed to have originated in Russia]." (CISA 2019a)

Lastly, the threat of distributed denial of services (DDoS) are growing. In 2016, a DDoS attack was conducted against a French Internet provider and the attack activated over 150,000 malware that infected internet-connected cameras, resulting in them consuming over one terabit per second of network throughput, which then crippled the provider's network capabilities and clients' websites (McAfee 2017).

The aging infrastructure is a concern to the chemical sector as well as to many of the other fifteen sectors. The age of transportation systems could soon render most of the US critical infrastructure vulnerable to disruptions. The chemical sector is highly dependent upon the transportation sector to operate, and in their "2017 Infrastructure Report Card, the American Society of Civil Engineers, in their 2017, rated the U.S. infrastructure as a D+; roads received a D; bridges, a C+; ports, a C+; rail, a B; and inland waterways, a D" (American Society of Civil Engineers 2017).

Partner sectors

The chemical sector has two major dependencies with the transportation and energy sector. Clearly, the energy sector delivers much-needed energy to power the chemical facilities, and much of this power travels to facilities via power lines, but also in trucks utilizing the transportation sector. Products depart chemical facilities utilizing the transportation sector via roads, rail, and inland river craft and oceangoing ships.

Chapter 4
Commercial Facilities Sector

The commercial facilities sector includes multiple venues that the public would normally flock to before the COVID-19 global pandemic, which includes opportunities that rely heavily on the public, such as business engagements, shopping at malls, entertainment in theaters, attending sporting events in stadiums, and lodging. These commercial facilities are based on maintaining direct access to the public and affording them the freedom to move about. These facilities are considered soft targets, which means that there are minimal security personnel and processes, minimal security barriers, and the areas are trafficked heavily by people moving in, out, or through them. These facilities and the concentration of massed people would certainly be highly advantageous to terrorists. Due to the high-profile nature of these events, the DHS secretary can designate them as a National Special Security Events (NSSE), which affords them greater protective resources and support by federal agencies (CISA 2015a). The majority of commercial facilities are privately owned and operated, with "minimal interaction with the federal government and other regulatory entities," and are comprised of eight subsectors (CISA 2015a).

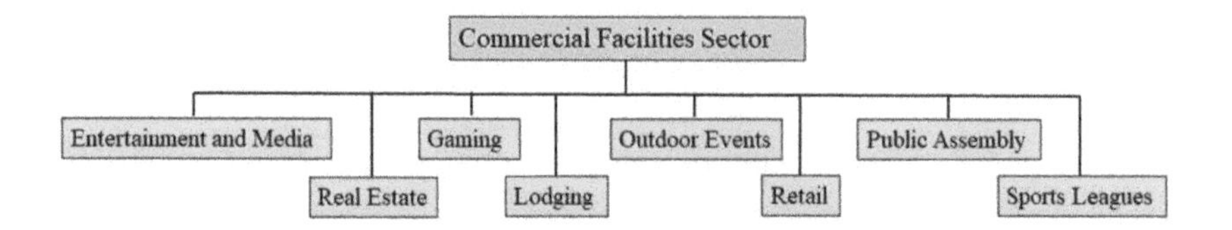

The US entertainment and media components are a

> "$703 billion market comprised of businesses that produce and distribute motion pictures, television programs and commercials, streaming content, music and audio recordings, broadcast, radio, book publishing, video games and supplementary services and products." (Westcott 2017).

Entertainment and media component

US entertainment and media are the largest globally and have captured 33 percent of the global market. Estimates are that the US entertainment and media subsector grew to $804 billion at the end of 2021 (Price Waterhouse Coopers 2018). The two major categories in this subsector are film and book publishing. The US film industry is comprised of "films, movie theaters, TV subscriptions, streaming content, and the distribution of film entertainment" (Navarro 2021), with $11.32 billion in box office sales in 2017 and $107.9 billion in home video sales in 2017.

Publishing component

The US publishing subsector includes paper and digital books in three major segments: "professional, educational, and consumer publishing" (Watson 2021). Consumer books have the largest market share, followed by educational and professional books. The reading format demanded by today's consumers has shifted from paper books to digital ones, such as e-books and audiobooks. This category is the largest globally, estimated to be "$116 billion; it had $37 billion in sales" (Watson 2021).

Gaming component

The gaming subsector consists of 462 commercial and 524 tribal casinos, over half of the global total (Yakowicz 2021; Lock 2020). In 2015, the gaming industry contributed "$240 billion to the U.S. economy, supported more than 1.7 million jobs, provided nearly $74 billion in income to the casinos, and generated $38 billion in tax revenues to FSLTT governments" (American Gaming Association 2015). According to the National Council on Problem Gambling, "approximately 85 percent of U.S. citizens have gambled at least once in their lives" (American Gambling Association 2015).

Outdoor events component

The outdoor events subsector is comprised of "fairs, exhibits, outdoor venues, parades, and 564 amusement and theme parks which attracted 290 million visitors to the parks alone in 2010" (International Association of Amusement Parks and Attractions 2015). In 2020, the outdoor recreation economy contributed "$374.3 billion to the U.S. gross domestic product (GDP)" (US Bureau of Economic Analysis 2020). Each year, over 3,200 fairs are held in North America; state fairs often attract one million visitors; amusement parks generate "$12 billion in revenues" (International Association of Amusement Parks and Attractions 2015). Large gatherings such as Macy's annual Thanksgiving Day Parade "can attract more than 3.5 million people to New York City" (Durando 2015).

Public assembly component

The public assembly subsector consists of "124,773 stadiums, arenas, theaters, museums, zoos, libraries, and other performance venues" where many people can congregate (US Census Bureau 2015a). This subsector significantly affects the economy as museums "support more than 726,000 American jobs and contributed $50 billion to the U.S. economy in 2016" (Stein 2018). Museums generate more than "$12 billion in tax revenue, one-third of it going to state and local governments; each job created by the museum sector results in $16,495 in additional tax revenue" (Stein 2018). Museums and other nonprofit cultural organizations "return more than $5 in tax revenues for every $1 they receive in funding from all levels of government" (Americans for the Arts, n.d.).

Real estate component

The real estate subsector includes "one million office buildings with over 9.9 billion square feet of office space and over 18 million multi-family households" (BOMA 2012; National Multifamily Housing Council, n.d.) across the US. Office buildings alone contributed "$205.1 million to the U.S. GDP annually" (BOMA 2012). In the US, the commercial real estate market represents "13 percent of its GDP by revenue and generates or supports nine million jobs" (The Real Estate Roundtable 2009). In addition, there are approximately "48,500 self-storage facilities and 4,000 facilities where self-storage serves as a secondary source of revenue" (The Real Estate Roundtable 2009).

Lodging component

The lodging subsector consists of "52,887 hotel properties which generate $163 billion annually" (American Hotel & Lodging Association 2014). This subsector includes "non-gaming resorts, hotels and motels, hotel-based conference centers, and bed-and-breakfast establishments" (US Travel Association 2014). Travel and tourism are the "sixth-largest employer in the U.S. and generated $2.1 trillion in 2013" (US Travel Association 2014).

Retail component

The retail subsector consists of "1.1 million buildings, including malls, shopping centers, and retail establishments, which generates $2.5 trillion to the U.S. GDP annually" (US Census Bureau 2015a; National Retail Federation 2013). This subsector is the "largest private-sector employer in the U.S., accounting for one in four U.S. jobs, totaling about 42 million positions and making up one-fifth of the total U.S. economy" (National Retail Federation 2013). Nationwide, online shopping reached "$370 billion in 2017" (Forrester Research Inc. 2015).

Sports league component

Lastly, this sector consists of major sports leagues and federations with "4,000 outlets and supports more than 133,000 jobs (US Department of Labor 2013; US Census Bureau 2015a). The sports league subsector attracted 134 million people to sporting events of the top "four major sports leagues alone—National Football League, National Basketball Association, National Hockey League, and Major League Baseball—and produced $23 billion in revenue" (Plunket Research 2015). The sports league subsector attracted "134 million people to only the top four sporting events in the U.S.—baseball, basketball, football, and hockey—and is estimated to be $485 billion in size" (ESPN 2014, Plunket Research 2015).

Partner sectors

This critical infrastructure sector has multiple interdependencies with many other sectors to function. Energy, communications, and water are required for any public assembly event; in the case of a sporting event in a stadium or a performance venue in an arena, the crowd size could number 100,000; according to the mayor of Rio de Janeiro, his city had "1.17 million tourists, over 400,000 were foreigners, during the 2016 Olympic Games" (Kalvapalle 2016). So for a relatively small public assembly, say a college football bowl game or an NFL playoff game, an assembly of 100,000 fans would be dependent on at least nine sectors of US critical infrastructure (CISA 2019a). The sectors of energy for power, communications, financial systems to process business transactions, emergency services to react to human incidents, food, and agriculture to provide food and beverages, water that is potable, and transportation sectors to move people and workers, and finally, nearby government facilities would manage and oversee security challenges. All would be required to mitigate risk and host such an event.

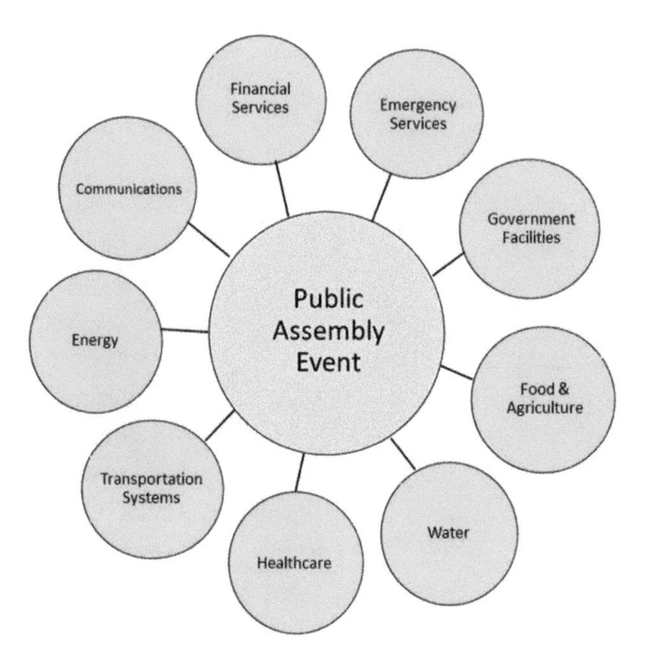

Threats

The threat from terrorism is always changing as bad actors find new or existing vulnerabilities to exploit and attack them. The 2013 Boston Marathon bombing was a public event that two homegrown extremists targeted. Insider threats references "individuals who work at a facility and use their inside knowledge to exploit and attack those vulnerabilities (CISA 2019a). As the diagram above demonstrates, there is a growing interdependency between multiple critical infrastructure sectors. If one fails due to cyber or a physical attack, it could cascade and impact the event or region that they serve. As discussed in chapter 19, our "adversaries have executed point-of-sale attacks on large retailers and hotels to gain access to confidential data, which has cost companies and financial institutions hundreds of millions of dollars" (CISA 2019a). There has also been a dramatic increase in "hacktivism," or politically or socially motivated cyberattacks. The facilities and building management systems, which control heating, ventilation, and air-conditioning systems, are computerized, making them vulnerable to cyberattacks or information technology (IT) outages. The critical infrastructure is dependent on the internet and IT; thus, the introduction of cyber tools and their subsequent failure would significantly affect the nation's safety, security, and economic well-being. The increased nefarious use of UAS and the absence of authorized C-UAS systems in the hands of trained security personnel at the thousands of critical infrastructure facilities is a significant threat that needs to be addressed. These systems can be used to cause damage to persons, property, and to critical infrastructure. UAS now allow bad actors an uncomplicated way to conduct surveillance, attack planning, and attacks on previously inaccessible areas.

Chapter 5
Communications Sector

The communications sector provides products and services that are critical to the operations of today's globally connected, "information-based society that supports 4.7 million jobs, contributes $475 billion to the U.S. GDP, and generates $1 trillion in economic output" (Accenture Strategy, n.d.; CISA 2015b). In the US alone, there are more than "262 million smartphones," and as a result, in the last two years alone, there has been a "238 percent increase in data traffic" (CTIA, n.d.a; CTIA, n.d.b; CTIA 2017).

Communication networks involve physical infrastructures, such as "buildings, switches, towers, antennas, and cyberinfrastructures, such as routing and switching software, operational support systems, and user applications" (CISA 2015b). Nearly all modern-day products are built with Internet connectivity via either an internal Bluetooth or Wi-Fi chip. Bluetooth enabled devices use short-range radio waves to connect to nearby devices and via a computer chip that broadcasts a signal; Wi-Fi also uses radio waves that also connects devices that are near one another, but it connects directly to the Internet, usually through a wireless router (Iscrupe 2020). The US is a highly Internet-connected and dependent nation; its economic and national security rely on a secure communications sector infrastructure to support operations, and it is the responsibility of the FSLTT and the private sector to work together to secure it. Today, massive amounts of information move at even faster speeds among an ever-larger number of users and machines. The communications sector consists of many facilities with unique functions, sizes, operating principles, and security risks. The communications sector architecture model (below) represents the collective infrastructure, which illustrates at least five major ways to access the numerous voice, video, and data services on the core network: broadcasting, cable, satellite, wireless, and wire line networks.

Broadcast

The broadcasting component has free and fee-based services such as radio and television, which can be analog or digital, audio, data, and video programming services. Broadcasting has been the "principal means of providing emergency alert services to the American public for the past six decades" (CISA 2015b). In the US, broadcasting systems operate in "three frequency bands: medium frequency for AM radio, remarkably high frequency for FM radio and television and ultra-high frequency for television" (CISA 2015b). The transition to "digital television and digital radio provides broadcast television and digital radio

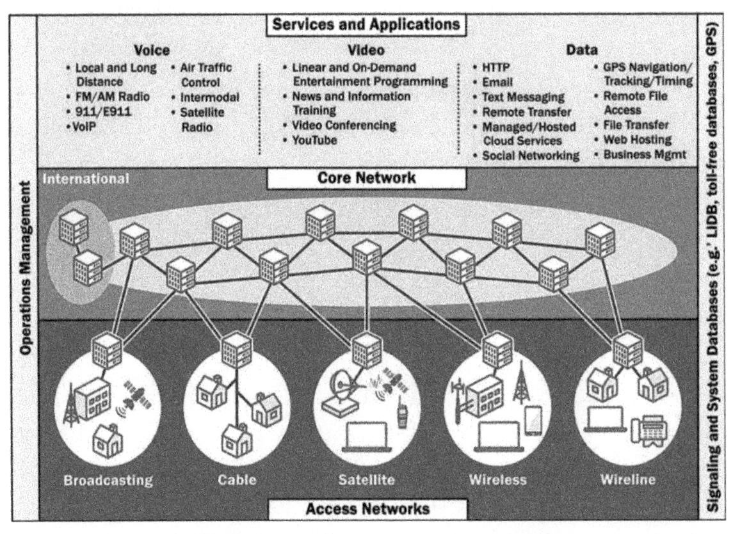

Image courtesy of CISA, Introduction to the Communications Sector Risk Management Agency, https://www. cisa.gov/sites/default/files/ publications/Communications%20SRMA%20Fact%20 Sheet_508.pdf

stations with the ability to multiple multicast programs on a single channel, and they can also stream broadcast and additional programming content over the Internet" (CISA 2015b).

Cable

The cable component consists of "more than 7,700 cable systems that offer analog and digital video programming services, digital telephone service, and high-speed broadband services" (CISA 2015b). The cable systems are dependent on a combination of fiber and coaxial cable to provide two-way signal paths to customers. This architecture benefits customers and businesses because it provides consistent signal performance, bandwidth, and network reliability.

Satellite

The satellite component is an integral part of the telecommunications network and consists of space-based platforms that are in geosynchronous orbit to relay voice, video, or data signals. Satellites in orbit still require earth stations to operate for "the two-way transmission of voice, video, and data services; data collection; event detection; timing; and navigation" (CISA 2015b). Multiple earth stations transmit signals to the satellite, which are then amplified and redirected back to another location on earth, or they can bounce between multiple satellites around the earth before being directed back to terrestrial stations.

Wireless

The wireless component is named after how the signal is transmitted over the communications path via electromagnetic waves instead of wire. Wireless technologies are found in "cellular phones, wireless hot spots, personal communication services, high-frequency radio, unlicensed wireless, and other commercial and private radio services to provide communication services" (CISA 2015b).

Wire line

The wireline component consists of networks with circuit switches and packet switches with copper, fiber, and coaxial cables to allow voice and data transmission. This remains the backbone of the Internet between continents.

Threats to the communication sector

Several hazards to the communication sector are equally dangerous to the other fifteen critical infrastructure sectors. These hazards include natural disasters such as hurricanes, wildfires, solar flares, aging infrastructure, pandemics, and high winds; cyberattacks; the use of UAS to conduct surveillance and attack. As we have witnessed thus far at the end of 2021, supply chain vulnerabilities are equally disastrous. All sixteen sectors depend on suppliers for the products, replacements, and services necessary to continue the smooth operation of the facilities. The communication sector is highly dependent on reliable hardware and software, but communications are global, and there are many partners, suppliers, facilities outside the country that could provide replacements or entirely new architecture in key locations if the supply chain is functioning properly. The global Internet consists of many suppliers, networks, and service providers, and a vulnerability in one can quickly migrate to other networks in countries worldwide. We can only stay ahead of the threat through constant vigilance from those wishing us harm through cyberattacks.

The DHS Quadrennial Homeland Security Review lists several what it terms long-range strategic threats, which include emerging risks by China and Russia to the nation's Global Positioning System (GPS), as well as increasing threats to the IoT, and an ability to rapidly mobilize assets, resources, and coordination risks associated with the rapid mobilization and coordination of all sixteen sectors of critical infrastructure in response to a large-scale, man-made, or natural threat to national security.

Partner sectors

According to CISA, the key cross-sector dependencies and interdependencies reside with the communications, energy, transportation, and water sectors. This multiple sector interdependences are essential to the daily operations of all sixteen sectors of critical infrastructure and in providing goods and services for the public. The communications sector is dependent on a few other sectors in particular energy, but the other fifteen sectors of critical infrastructure are all dependent on communication products and services for their daily operations and services, such as organizational networks, Internet connectivity, voice services, and video teleconferencing capabilities (CISA 2015b). More specifically, the emergency services sector "relies on networks for emergency operations center connectivity, interconnecting land mobile radio networks, backhauling traffic, operating public alert and warning systems, and receiving emergency 911 calls"; the energy sector relies on the communications sector to monitor and control its operations and electric transmissions; the financial services sector relies on the communications sector for "transmitting transactions and financial market operations"; the IT sector is dependent on the communications sector for network delivery and distribution of its applications and services; lastly, the transportation sector relies on the communications sector to monitor and control the transportation infrastructure, such as "traffic signals, mass transit, air traffic control, and vehicle traffic monitoring operations" (CISA 2015b).

The communications sector has multiple dependencies. Communication networks require electricity to operate, thus there is a dependency upon the energy sector to include reliance on diesel fuel to provide the fuel to power backup generators and the transportation sector to deliver fuel. There is a dependency upon the water services sector for water for consumption, cooling processes, and for wastewater operations. The communications sector is also dependent on IT "to deliver reliable products, such as routers, switches, software, operating systems, and services such as domain name resolution, to provide end-to-end communication services for customers" (CISA

2015b). The communications sector is surprisingly dependent on the defense industrial base sector principally for the use of its Global Positioning System (GPS), a part of the defense industrial base sector that is required for precision timing and network synchronization functions (CISA 2015b). Many other critical infrastructure sectors are dependent on the communications sector. As such, it is one of the few sectors upon which all others depend. Each of the other sectors is dependent communication services that this sector provides and is necessary to support their respective operations and "associated day-to-day communication needs for corporate and organizational networks and services such as Internet connectivity, voice services, and video teleconferencing capabilities" (CISA 2015b).

Chapter 6
Critical Manufacturing Sector

The critical manufacturing sector "processes raw materials and produces highly specialized essential parts and equipment" for transportation, defense, electricity, and major construction (CISA 2015c; CISA 2019c). In 2020, the manufacturing sector contributed "$2.7 trillion to the U.S. GDP, including direct and indirect purchases from other industries; value-added manufacturing contributed an estimated 24 % of GDP" (US Department of Commerce 2021). In 2019, the manufacturing sector "imported 14.7 % of its intermediate goods, resulting in 10.7 % of the output being of foreign origin" and employed "15.7 million people, representing 10% of the total U.S. employment" (US Department of Commerce 2021). By itself, the US manufacturing sector ranks as the "sixth-largest economy globally and produces 11.39% of the economy's total output" (Silver 2021). Many of the products created by the manufacturing sector are critical to the energy and the defense industry; thus, a disruption in manufacturing would ripple across multiple other sectors of critical infrastructure and could impact the nation's national security. Many manufacturing sector products are integrated into the highly interdependent global supply chains.

Globally, nearly all companies that manufacture products today are "interconnected through a chain of suppliers, vendors, partner companies, integrators, contractors, and customers that link them all to other businesses" (CISA, 2019c) in other critical infrastructure sectors. The manufacturing sector is concentrated largely around major ports as the geographic concentration provides readily available expertise, reduced logistics costs, the rapid delivery of imported raw materials, and the export of products internationally. The international networks of raw materials and finished products are complex, fickle depending on geopolitical conditions, and exasperate risk for US operations.

The image courtesy of CISA, the Critical Manufacturing Snapshot, https://www.cisa.gov/sites/default /files/publications/nipp-ssp-critical-manufacturing-2015-508.pdf; National Association of Manufacturing, Facts About Manufacturing. http://www.census.gov/econ/ snapshots /index.php; Rosselot, Jason. Partnership for More Secure Manufacturing. Panel: Industrial Cyber Security: Critical Manufacturing Sector Coordinating Council. April 2015. http://www.nam.org/Statistics-And-Data/Facts-About-Manufacturing/Landing.aspx

The four major components of the manufacturing sector "include primary metals manufacturing, machinery manufacturing, electrical equipment appliance and component manufacturing, and transportation manufacturing" (US Census Bureau 2015b). In 2012:

> "4,556 primary metals manufacturers received raw materials and converted them into assemblies, intermediate products, and end products which included metal sheets, bars, beams, slabs, or pipes worth $270.9 billion in U.S. shipments." (US Census Bureau 2015b)

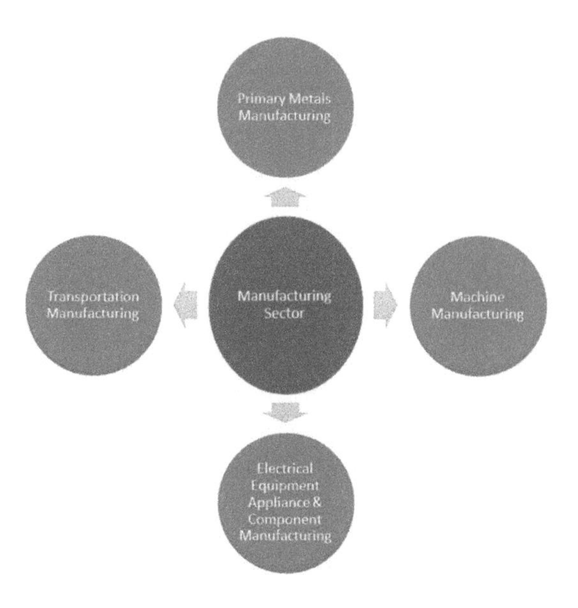

The machinery manufacturing component produces "engines, turbines, and power transmission equipment," which are key to the operations in several other critical sectors of critical infrastructure. There are "24,124 heavy equipment manufacturers such as earth-movers, mining, agriculture, construction, and other heavy material handling equipment, which was responsible for $407.6 billion in U.S. shipments in 2012" (US Census Bureau 2015c).

The "electrical equipment, appliance, and component manufacturing components produce specialized equipment, assemblies, intermediate products, and end products for power generation, including transformers, electric motors and generators, and industrial controls (US Census Bureau 2015b). In 2012, there were "5,765 manufacturers whose products resulted in $123.9 billion in U.S. shipments" (US Census Bureau 2015d). In 2012, the transportation manufacturing component "Had 11,814 manufacturers of cars and trucks, aircraft and parts, aerospace products and parts, railroad cars and other railroad products, and other transportation equipment and was responsible for $792.9 billion in U.S. shipments." (US Census Bureau 2015e)

Threats to the critical manufacturing sector

Risks to the critical manufacturing sector include natural disasters, terrorism, and cyberattacks. Many of this sector's facilities are in regions synonymous with extreme weather anomalies, such as hurricanes, increasing their vulnerability, yet many of the sector's facilities and suppliers dispersed globally. Thus, this arrangement could offset focused disruptions in a country that would not necessarily cascade around the globe. However, extremes in weather could disrupt the energy, water, and transportation sectors, which would cause subsequent delays in moving manufactured goods. While US supply chains are highly efficient, they are not redundant. Years ago, businesses moved to a just-in-time model for improving overhead costs, but the limited inventories have inadvertently made key manufacturers in this sector sensitive to interruptions in the supply chain processes, and their operations will be disrupted if the raw materials that they depend on do not reach the manufacturing facilities in a steady fashion. Customers, in turn, would be affected by an inconsistent delivery of finished goods. Thus, this sector's manufactures must constantly monitor geopolitical, supply chain, and other factors to determine what actions must be taken to offset them.

The strategic and economic importance of such a sector makes it an ideal target for terrorists' intent on destroying facilities or interfering with supplies as an attack might result in mass casualties and would certainly garner media attention as it did with the 2019 UAS and cruise missile attack on Saudi Arabia's number one critical infrastructure (see chapter 20). Manufacturing processes typically use "commercially available software and standard industrial control systems rather than proprietary system designs" (CISA 2019c). Cyber intruders would likely attempt to seize control of these systems to disrupt processes, corrupt information, intentionally damage equipment, or steal proprietary information. The theft of intellectual property via cyberattacks has been on the rise for the past few years and can reduce competitiveness and destroy reputations.

Partner sectors

The critical manufacturing sector is thoroughly integrated with other critical infrastructure sectors, which has created interdependences such that disruptions in one sector would certainly affect operations in another. Manufacturers require constant, uninterrupted power for operations, and a short- or long-term loss of power would significantly disrupt their operations. In many facilities, a continuous water source is essential for manufacturing processes and is certainly required for human operators. The communications sector enables "coordination of supply chain movements, control system processes, emergency notification and response" (CISA 2019b). The transportation sector enables the movement of materials and products around the globe, often on strict timelines, using aviation, freight rail, highway, and maritime means. Even in the twenty-first century, the transportation sector is at risk of shipping vessel piracy and limited availability of rail lines. The critical manufacturing sector is also dependent on the information technology sector for its "manufacturing operations, global transit, quality control systems, critical processes, and facility security" (CISA 2019b).

Chapter 7
Dams Sector

The dams sector performs two significant services: critical water retention and water control. The sector responsibilities include

> "hydroelectric power generation, municipal and industrial water supplies, agricultural irrigation, sediment and flood control, river navigation for inland bulk shipping, industrial waste management, and recreation." (CISA, n.d.)

The dams sector provides "irrigation for over 10 percent of U.S. cropland; protects more than 43 percent of the U.S. population from flooding; and generates 60 percent of electricity in the Pacific Northwest" (CISA, n.d.; CISA 2015d). There are more than "90,000 dams in the U.S., and 65 percent of them are privately owned and operated" (CISA, n.d.; CISA 2015d). The dams sector directly supports several critical infrastructure sectors and various industries.

Partner sectors

The critical infrastructure interdependencies includes the communication, energy, food and agriculture, transportation, and water sectors. The communications networks "enable remote dams sector operations and control (CISA 2015d). The energy sector is dependent on numerous hydropower dams to provide electricity and "black start" capabilities, which is a process of, "Restoring power to and operations of an electric power station or part of an electrical grid to operation without relying on the external electric power transmission network to recover from a total or partial shutdown." (CISA 2015d)

Normally, the electric power used within the plant is provided by the station's generators. The food and agriculture sector are dependent on dams to "provide water for irrigation and protect farmland from flooding" (CISA 2015d). Transportation systems use navigation lock systems in the dam's sector to aid all inland and intracoastal waterway freight movements. Additionally, major roadways may also traverse dams. The dams sector also provides "drinking water supplies and pumping capabilities for the water sector" (CISA 2015d).

Image courtesy of https://www.water-technology.net/projects/grand-coulee-dam-washington-us/

One of the largest dams and the greatest electricity producer within the US is the Grand Coulee Dam, a concrete gravity dam on the Columbia River in Washington. Work began in 1933 was completed in 1942 to produce hydroelectric power and water for irrigation. Grand Coulee originally had two powerhouses, but in 1974, a third was added, increasing its energy production. Through a series of "upgrades and the installation of pump generators, the dam now supplies four power stations; the dam's reservoir supplies water for the irrigation of 671,000 acres (DOE 2010). Grand Coulee Dam is 550 feet tall; 5223 feet long; has a width of 30 feet at the crest and 500 feet at its base" (DOE 2010). Its annual electricity production is "6,809 megawatts, equivalent to 21 billion kilowatt-hours" (DOE 2010).

For comparison, the Three Gorges Dam in Hubei, China, is the number one dam producer of electricity in the world as it generates as much as 22,500 megawatts depending on the annual precipitation in the river basin (Cleveland and Morris 2013; Ehrlich 2013). One thousand megawatts equal one terawatt.

Threats to the dams sector

> "Protecting the nation's dams' sector critical infrastructure, which provides a wide range of economic, environmental, and social benefits, is essential to maintaining a resilient nation. Because the vast majority of the sector's assets are privately owned and operated, active collaboration and information sharing between the public and private sector partners are essential to increasing the security and resilience of the nation's dam's sector's critical infrastructure. Dams, levees, and related facilities are vital to the nation's infrastructure, providing a wide range of economic, environmental, and social benefits through delivering critical water retention and control services. Therefore, incidents involving dams sector assets could result in severe economic losses, loss of life, and reduced public confidence in industry and government's ability to provide essential services." (CISA, n.d.)

The threats to the dam sector are both natural and man-made. Natural disasters are often listed in the critical infrastructure sectors as a threat, but in the case of the dam's sector, their impact cannot be discounted as the natural disasters can lead or accelerate structural issues from internal and external erosion. Terrorism attacks focused on physical assets or even on the dams themselves and are a possibility that requires constant surveillance and other electronic devices used to detect certain explosives or chemical, biological, and radiological devices. Due to the dams sector's reliance on IT, the threat of a cyber-attack remains a persistent risk. Owners and operators depend upon modern control systems, many of which have commercial hardware and software systems, and many have transitioned to remote monitoring and control processes" (CISA 2015d), thus the risk of cyber-terrorism or of a cyberattack has only increased.

Droughts reduce water tables; severe storms increase water levels, and both can create strains on the dam's internal infrastructure. The aging of multiple sectors is a concern that needs to be addressed by our elected officials. It is a risk that simply does not get any better with the passage of time. In fact, it only worsens. It is estimated that there are an estimated "4,000 maintenance deficient dams, as is 91 percent of federally inspected levees" (CISA 2015d). Lastly, due to the growth of human populations and the subsequent housing and business developments near dams and levees, the hazard potential of this sector is quite naturally going to increase due to the increased risk of asset failure.

Chapter 8

Defense Industrial Base Sector

The US stands atop the global community with its defense budget dwarfing that of the next seven countries as "$768 billion in fiscal year 2021" (Edmondson 2021). The total global military expenditures were approximately, "$1.917 trillion in 2019 and the ten countries with the highest defense spending were the following: U.S. at $768 billion; China at $237 billion; Saudi Arabia at $67.6 billion; India at $61 billion; United Kingdom at $55.1 billion; Germany at $50 billion; Japan at $49 billion; Russia at $48 billion; South Korea at $44 billion; and France at $41.5 billion." (World Population Review 2021)

The US national security expenditures are the "third most expensive government program behind Social Security and Medicare in the fiscal year 2018 budget where the U.S. spent $4.11 trillion" (Congressional Budget Office 2018).

The defense industrial base refers to the "nation's industrial assets that are of direct or indirect importance to the production of equipment and technologies for use by a nation's military" (CISA 2010). The US defense industrial base sector consists of "over 250,000 organizations from across the federal government, private sector, and worldwide who aid in accomplishing the following: equip, inform, mobilize, deploy, sustain and support military operations directly; perform research and design; design, manufacture, and integrate systems; and maintenance depots and service military weapon systems, subsystems, components, subcomponents, or parts, all of which are intended to satisfy U.S. military, national defense requirements." (CISA 2010; Looking Glass, n.d.)

The supply chain for the quarter of a million organizations tied to the defense industrial base is extraordinarily complex. Since it is intricately woven with national security, our strategic competitors have realized that targeting vulnerable companies involved in the defense industrial base supply chain can be a vulnerability that they can exploit and profit from. The government component of the defense industrial base consists of the following: "laboratories, special-purpose manufacturing facilities, capabilities for production of uniquely military equipment such as arsenals and ammunition plants, and other services" (CISA 2010). The defense industrial base companies can also "deliver national security products and services to other federal agencies," such as air defense or C-UAS if needed (CISA 2010). The immense size, number of organizations, vast resources, and budget expenditures of the defense industrial base sector does include the infrastructure of other sectors of critical infrastructure such as power, transportation, or communications, all of whom the DoD is dependent on to execute military operations.

Industry Segments	
Industry Segments	Industry Sub segment
Aircraft	Fixed Wing
	Rotary Wing
	Unmanned Aerial Systems
Ships	Surface
	Sub-Surface
	Unmanned Underwater Vehicles
Tracked and Wheeled Land Vehicles	Combat Vehicles
	Tactical Vehicles
	Unmanned Ground Vehicles
Electronics	Electronic Warfare
	Command, Control, Communications, Computer and Intelligence (C4I)
	Avionics
Soldier Systems	Chemical Biological Defense Systems
	Clothing and Textiles
	Subsistence/Medical
Structural	Castings/Forgings
	Composites
	Armor (Ceramic/Plating)
	Precious Metals
Munitions	Missile Tactical
	Missile Strategic
	Missile Air/Air
	Missile Air/Surface
	Missile Defense
	Missile Surface/Air
	Missile Surface/Surface
	Precision Guided Munitions
	Ammunition
	Missile Defense Agency
Space	Launch Vehicles
	Satellite
	Missile Defense Agency
Mechanical	Transmissions (Air/Auto)
	Propulsion (Diesel/Rocket/ Turbine)
	Hydraulics
	Bearings
	Nuclear Components (includes Depleted Uranium)

Image courtesy of CISA, The Defense Industrial Base Sector-Specific Plan, https://www.cisa.gov/sites/default/files/publications/nipp-ssp-defense-industrial-base-2010-508.pdf

The former undersecretary of defense for acquisition and sustainment, Ellen Lord, recently stated that "over a period of years, we have offshored many, many sources of supply. It's not for one reason; it's for a variety of reasons, whether it be regulations, whether it be labor costs, whether it be government support of different industries" (Lopez 2021). According to Ms. Lord:

> "the deindustrialization of the U.S. over the last 50 years, the end of the Cold War and the focus it gave the U.S. on defeating the Soviet Union,

digital technology and the rise of China have all created challenges to national defense. There are a couple key areas there with shipbuilding, as well as microelectronics, fundamental to our capability. Development of a modern manufacturing and engineering workforce along with a more robust research and development base is also critical. Declines in U.S. science, technology, engineering and mathematics education and industrial jobs hurt the ability of the defense industrial base to innovate. We want to make sure that we have modern manufacturing and engineering expertise. We do not have nearly the number of scientists and engineers as China has. We need to make sure that we develop our talent to be able to leverage on these critical areas. The department must also reform and modernize the defense acquisition process to better meet the realities of the 21st century, Lord said. We've started with a number of those, but there's much further to go. We want to make sure that our traditional defense industrial base is widened to get all of those creative, innovative companies. We know the small companies are where most of our innovation comes from, and the barriers to entry, sometimes to getting into the Department of Defense, are rather onerous. Part of modernizing and reforming defense acquisition is the recently announced Trusted Capital Marketplace, which will match potential defense suppliers, many of them small companies that have never done business with DOD, with the investors they need to keep operating and innovating. The Trusted Capital Marketplace will vet investors to ensure foreign ownership, control and influence are nonexistent." (Lopez 2021)

Threats to the defense industrial base

The image below denotes several areas of national security concern for the US, such as the reliance on foreign shipping vessels to transport "97 percent of imported products to our shores, and China's dominance in shipping fleets and containers, the ownership of shipyards, production of cranes, microelectronics, space satellites, and UAS manufacturing" (Lopez 2021). The US national security vis-à-vis the defense industrial base is dependent on foreign parts manufacturers and on foreign shipping companies. In the event of a major conflict with Russia or China, it is likely that all means of warfare would be exerted by our strategic competitors, to include withholding rare earth minerals from China, and economic warfare to include withholding other supplies which our nation's military and economy is highly dependent on. Under Chinese president Xi, everything that China does is focused on increasing the global dependency on China while increasing its military to execute any missions given to it by its leader. Article 7 of China's National Intelligence Law passed in 2017 "requires organizations to assist in espionage and to keep those activities secret" (Girard 2017). Everything that Chinese government officials and private companies do in the global marketplace is shared with the central government and the military hierarchy to further Chinese interests. Inadvertently, the laissez-faire attitude and celebration of the US leadership in the post-Soviet era has only weakened the national security of the US.

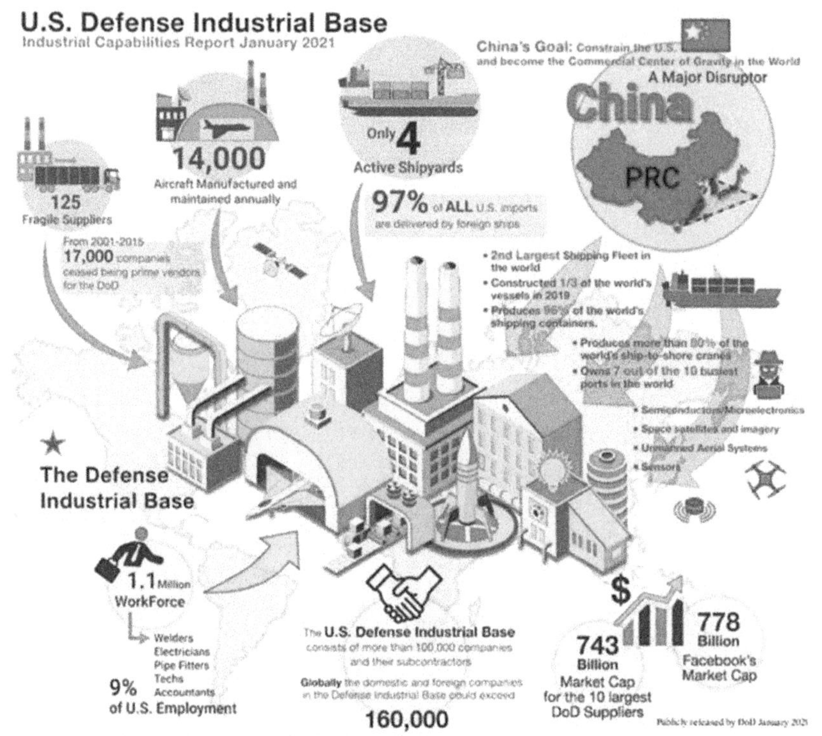

Image courtesy of Lopez, 2021; DoD News; https://www.defense.gov/News/News-Stories/Article/Article/2474015/dod-aims-to-bring-industrial-base-back-to-us-allies/

Partner sectors

The defense industrial base does have many connections with DHS for overlapping threats from overseas originating overseas to the homeland. The defense industrial base underwriting of GPS has direct impacts upon multiple sectors including energy, critical manufacturing, communications, commerce, and information technology for "precision timing, location, and IT network synchronization functions" (CISA 2015b). The energy sector has an interdependency with the defense industrial base due to the military and civilian applications involving nuclear energy while reducing the threat of global terrorism and the use of WMDs. The commercial facilities sector aid in the advancement of "U.S. national security, foreign policy, and economic objectives by ensuring an effective export control and treaty compliance system and promoting continued U.S. strategic technology" (CISA 2010). The information technology sector supports the defense industrial base as it is responsible for aiding in the deterrence, prevention, and defeat of cyberattacks.

Chapter 9

Emergency Services Sector

The emergency services sector encompasses a diverse arrangement of multi-disciplines and unique capabilities that empowers an extensive variety of services to include prevention, preparedness, response, and recovery to serve the public. In doing so, this sector also aids in the protection of the nation's critical infrastructure. The nation's emergency services operators work the front lines for nearly all emergency scenarios involving the sixteen sectors of US critical infrastructure. Clearly, disrupting the emergency services sector could negatively impact society through the infliction of injuries and death of citizens, promulgating major public health issues resulting in a short- to long-term loss to the economy, and a multitude of other disruptions that could cascade to the other sectors of critical infrastructure. Due to the criticality of this sector's importance, both real and perceived, the performance must be perfect each time it is needed, otherwise it would have an immediate and highly negative impact on the public's perception and morale and on the other sectors of critical infrastructure.

The responsibilities for recruiting, training, organization, and management are handled at the state, local, tribal, and territorial (SLTT) level, therefore there can be differences in skills, competence, and reaction times. The emergency service sector consists of highly trained, well-disciplined personnel prepared to react to wide-ranging emergencies from the following disciplines: "law enforcement; fire and rescue services; emergency medical services; emergency management; and public works" (CISA 2015e). The wide variety of required missions and the assets needed to accomplish those tasks makes the emergency services sector unique among the sixteen critical infrastructure sectors and an inability to execute its mission would immediately negatively impact the other sectors. The emergency services sector is comprised of systems and networks in the physical, cyber, and human components.

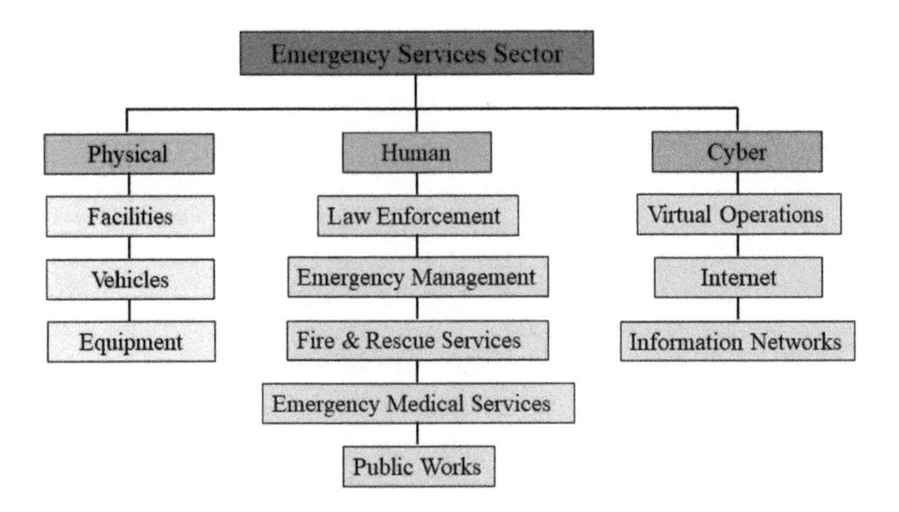

Physical

The physical component includes "operations, storage facilities, specialized equipment, and vehicle fleets that enable personnel in each component of the emergency services sector to perform critical services during steady-state and crisis operations" (CISA 2015e). The emergency services sector requires a specialized vehicle fleet, equipment, and communications systems which entail additional training, maintenance, and storage requirements.

Cyber

The emergency services sector operations are highly dependent on information technology, specifically the communication systems, databases, biometrics, telecommunications, and on the Internet for warning, information, alerts, and support, and electronic security systems. As such, all are performed virtually and are vulnerable to cyberattacks and disruptions. Any degradation of any of the aforementioned systems that emergency services personnel are dependent upon would increase risk to the public, to emergency service facilities, and to emergency responders and would negatively impact emergency operations.

Human

The component consists of "2.5 million career and volunteer specialists" (CISA 2015e) across the FSLTT realm who serve in communities throughout the US. Within the human component, there are five distinct disciplines: "law enforcement, fire and rescue services, emergency medical services, emergency management, and public works" (CISA 2015e). The focus of these five disciplines is to prepare for and execute response operations, protect residents and property, and ensure public order in times of disaster, thereby contributing to the safety and security of the nation. Law enforcement assets enforce laws, conduct criminal investigations, collect evidence, apprehend suspects, secure the judicial system, and ensure custody and rehabilitation of offenders. Law enforcement consists of police departments, sheriff's offices, courts systems, correctional institutions, and private security agencies. Fire and rescue services "perform fire suppression, fire prevention, hazardous materials control, life, and property safety operations, building code enforcement, and fire safety education and consist of paid and volunteer personnel" (CISA 2015e).

The emergency medical services are responsible to provide medical care at the scene of an incident such as triage and treatment; to provide medical services during the outbreak of an infectious disease to protect staff, patients, facilities, and the environment; and transporting ill and injured patients to medical facilities (CISA 2015e). The emergency management discipline provides coordination and incident management between the other four disciplines and with nonemergency service entities. Emergency operations centers provide for multiagency coordination for incident management and the "coordination of response and recovery activities among neighboring jurisdictions at various levels" (CISA 2015e), plus FSLTT if warranted. The public works discipline is responsible to assess and repair damage to buildings, roads, and bridges; clearing debris from public spaces; restoring utility services; managing emergency traffic; hardening security enhancements to critical facilities; and monitoring the safety of public water supplies. Public works departments often supply heavy machinery, raw materials, and emergency operators. They may also be responsible to manage contracts for additional labor, equipment, or services that may be needed before, during, and after an emergency, thus they are an integral component of a jurisdiction's emergency planning.

Notable trends and emerging issues

There are eight trends that are key to critical infrastructure sector partners as they adapt to security issues and build resilience to evolving risks. The first trend is to deal with increasing expectations by the public for the emergency services sector's "expertise, rapid response capabilities, and real-time information sharing" (CISA 2015e). Today, the emergency services sector's focus is on dealing with response planning for all possible hazards as threats continue to evolve. To counter public panic, the emergency services sector has designated personnel to provide real-time information to an information-hungry public.

A second trend is the reduction of grant funding which has strained both state and local resources. The reductions in budgets have affected the capacity of the emergency services sector to anticipate, prepare for, and respond to threats and incidents. Many communities have been forced to choose between increasing the ever-present and expensive everyday policing requirements or acting upon mitigation strategies to the threat posed by terrorists. Increased costs for personnel, health care, fuel, and maintenance of equipment have begun to limit the services available to separate or multiple incidents.

The third trend involves extreme weather events. If the frequency of extreme weather events increases, the response demands will increase, and these will drain from available sector personnel, assets, capabilities, resources, and the budgets of affected municipalities.

The fourth trend is a worldwide one which involves a dependence on cyber infrastructure. The emergency services sector is more dependent than ever before on cyber assets, networks, and other technological systems to conduct its mission to protect the public and respond to emergency incidents. The technological advancements such as the "Next-Generation 9-1-1, the transition towards cloud-based information systems, and the use of geospatial tools have allowed the emergency services sector to expand and improve operations" (CISA 2015e). However, with new technological advancements come new risks and vulnerabilities, such as "data encryption limitations, location accuracy gaps, global positioning system (GPS) disruptions, driver distractions, and user information privacy issues, challenge the sector's capability to respond to emergencies quickly and safely" (CISA 2015e).

A fifth trend is the dynamics of a changing population. The population is increasing and aging, which translates to an increased number of connected mobile devices and an increased frequency of calls for emergency medical services. An increase in the population has also resulted in an increase in vehicles on the roadways and a corresponding increase in vehicle and truck accidents, which has also resulted in an increased request for emergency services and hazardous material teams.

A sixth trend is the intentional targeting of emergency services personnel. First responders are at risk due to the nature of their work; however, they may become the opportunity targets when they respond to mass-casualty incidents, an active shooter, or an improvised explosive device when the intent of a bad actor is to garner media attention while inflicting as much damage and casualties and he or she can inflict on an unsuspecting public. These incidents will have a negative impact upon the physical and emotional health of first responders and will often shorten their careers.

A seventh trend involves an aging infrastructure throughout the US. This aging in an affliction that impacts the nation's electrical grids, water and waste systems, and roads and bridges. Increased incidents due to an aging infrastructure that is not being addressed by the federal government has increased the incidents that require emergency services personnel and can certainly impede their efforts to get to those who require assistance.

An eighth trend is a decrease in the emergency services sector work force and in their expertise. Retirements and losses due to mental and physical difficulties will not be filled for many years until the inexperienced practitioners gain experience and recruiting fills the vacancies.

Threats to the emergency services sector

The most significant risks to the emergency services sector arise from "cyberattacks and other disruptions; natural disasters and extreme weather; violent extremist and terrorist attacks; and chemical, biological, radiological, and nuclear incidents" (CISA 2015e). The emergency services sector has become "increasingly dependent on cyber-based infrastructure and operations, including emergency operations communications, data management, biometric activities, telecommunications (e.g., computer-aided dispatch), and electronic security systems" (CISA 2015e). This reliance has increased the potential for attack and attractiveness for bad actors to target the emergency services sector through denial of service and attacks on information systems. The increased dependence on the Internet and interconnectedness of this sector as well as the other fifteen sectors of critical infrastructure has increased the cybersecurity risks across the board.

Natural disasters and extreme weather severity have increased the need for emergency services personnel and resources, often for extended periods. The mutual support agreements with other municipalities, cities, and regions will strain resources and responsiveness across larger geographic areas that were initially affected by the storm.

Violent homegrown extremist and terrorist attacks have increased, and the public is now more than ever before cognizant of the present-day threat of terrorism and insider threats. Those who support or commit violence have increasingly targeted emergency services personnel, especially law enforcement, and soft targets. The threat of terrorism will remain for the foreseeable future, and through innovation, bad actors are utilizing an increasingly wide range of attack approaches, such as improvised explosive devices that are placed in an area with the public, vehicle-borne ones that drive into crowds or are parked near a large crowd, unmanned systems that can be used to transport and disperse weapons of mass destruction (WMD) or to drop explosives from a height above unsuspecting members of the public. Secondary explosive devices have been used on many occasions in war zones, but they are now being used in terrorist attacks in increased frequency. These attacks clearly are meant to target first responders and pose a significant threat to the emergency services sector. The Internet and social media "enable lone actors and terrorists to identify and interact with others and obtain the ideological and material support needed for acts of violence" (CISA 2015e).

We have been fortunate to not have suffered through a chemical, biological, radiological, or nuclear (CBRN) incident in the country, but how long our luck will hold out is anyone's guess. Every day in the US and across the globe, FSLTT and international law enforcement agencies have a dedicated pool of personnel and resources focused on countering the threat posed by WMDs. Emergency services personnel have responded to and treated victims of industrial or transportation incidents involving CBRN, but the affected areas and personnel has been relatively small. A terrorist attack utilizing WMDs would aim for maximum dispersal across a highly populated area. The intentional use of CBRN would spread rapidly, even faster than COVID-19 has, and would impact "numerous jurisdictions and greatly strain emergency service resources and impact the health and safety of large numbers of the public and responders" (CISA 2015e). CBRN incidents require a great deal of highly technical training and specialized equipment for emergency services personnel. In such an incident, it is entirely likely that the DoD would be asked to provide specially trained and experienced soldiers specializing in this deadly arena.

Sector interdependency

There is a significant risk that the degradation of the emergency services sector would immediately and negatively have on the other fifteen sectors of critical infrastructure, the FSLTT echelons of government, the private sector across the realm of all industries, and the safety, security, and morale of the public. There is a corresponding dependence by the emergency services sector on other critical infrastructure sectors that are integral to the safety and ability of emergency services personnel to function. Those other sectors are energy, transportation systems, communications, water energy, and information technology. The dependence on the energy sector is critical to all aspects of a modern society and certainly to the emergency services sector. The transportation systems sector provides a resilient transportation network necessary to respond to emergencies and "transport people, goods, and services to and from incident areas" (CISA 2015e). Public communications are required for the emergency services sector, such as through "an internal communications network, 9-1-1 services, or other public alerting and warning systems" (CISA 2015e). The water sector provides a clean and reliable water supply, both critical in firefighting and public works. Emergency services is "dependent on a variety of cyber-related assets, systems, and disciplines to conduct its many missions; thus, the loss of computer-aided dispatch services, the corruption or loss of confidentiality of critical information, or jammed surveillance capabilities could significantly disrupt the sector's capability to adequately protect the public and safely and quickly respond to emergencies" (CISA 2015e).

Lastly, there are internal dependencies within the emergency services sector that are of the utmost importance. Each discipline is highly dependent on the other disciplines for their operations and success, such as law enforcement officers protecting the fire department personnel and emergency medical services (EMS) personnel at the scene of an emergency, while the public works personnel clear debris from emergency routes to facilitate the access to the affected site for first responders.

Chapter 10
Energy Sector

A stable energy infrastructure fuels the nation and its economy while providing for the health and welfare of its citizens. The energy sector was identified in "Presidential Policy Directive 21 as uniquely critical because it provides an enabling function across all 16 sectors of critical infrastructure" (The White House 2013). According to CISA:

> "More than 80 percent of the country's energy infrastructure is owned by the private sector, supplying fuel to the transportation industry, electricity to households and businesses, and other sources of energy that are integral to economic growth and production across the nation." (CISA 2015f)

A modern, globally interconnected nation must possess an energy infrastructure that provides uninterrupted energy to the government, businesses, and its citizens. In the US, the three main suppliers of energy are derived from electricity, petroleum, and natural gas, and a disruption would cause irreparable harm to the national security, the economy, and to the public.

There are a "wide variety of energy supply alternatives and delivery mechanisms, energy infrastructure assets and systems are geographically dispersed, and the millions of miles of electricity lines and oil and natural gas pipelines and many other types of assets exist in all 50 states and territories [thus protecting the sheer vast numbers of facilities and assets in the energy sector is an impossibility]." (CISA 2015f)

In the 2003 Homeland Security Presidential Directive 7 (HSPD-7), President Bush further outlined the "U.S. Critical Infrastructure Identification, Prioritization, and Protection" (The White House 2003). In HSPD-7, the energy sector is defined as consisting of "thousands of electricity, oil, and natural gas assets that are geographically dispersed and connected by systems and networks" (The White House 2003).

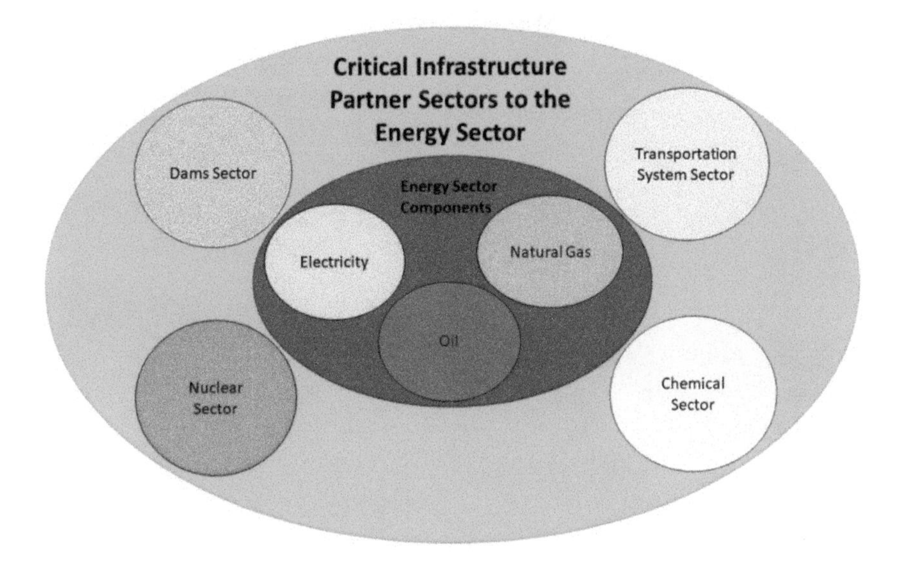

The energy sector is divided into "three interrelated subsectors: electricity, oil, and natural gas which includes the production, refining, storage, and distribution of oil, gas, and electric power, apart from hydroelectric and commercial nuclear power facilities and pipelines" (The White House 2003). The "2013 Presidential Policy Directive (PPD) 21, signed by President Obama, identified the energy sector as uniquely critical because it provides an essential function across virtually all critical infrastructure sectors" (The White House 2013).

Electricity	Petroleum	Natural Gas
• Generation – Fossil Fuel Power Plants » Coal » Natural Gas » Oil – Nuclear Power Plants[a] – Hydroelectric Dams[a] – Renewable Energy • Transmission – Substations – Lines – Control Centers • Distribution – Substations – Lines – Control Centers • Control Systems • Electricity Markets	• Crude Oil – Onshore Fields – Offshore Fields – Terminals – Transport (pipelines)[a] – Storage • Petroleum Processing Facilities – Refineries – Terminals – Transport (pipelines)[a] – Storage – Control Systems – Petroleum Markets	• Production – Onshore Fields – Offshore Fields • Processing • Transport (pipelines)[a] • Distribution (pipelines)[a] • Storage[b] • Liquefied Natural Gas Facilities[b] • Control Systems • Gas Markets

[a] Hydroelectric dams, nuclear facilities, rail, and pipeline transportation are covered in other SSPs.
[b] Certain infrastructure of this asset type are regulated by the Chemical Facility Anti-Terrorism Standards (CFATS). The final tiering of the facilities covered by the CFATS was not completed at the time of this report.

Image courtesy of CISA, Segments of the Energy Sector, https://www.cisa.gov/sites/default/files/publications/nipp-ssp-energy-2010-508.pdf

Energy Source	Billion of kWh		Share of Total	
Fossil Fuels	**2,427**		**60.6%**	
Natural Gas		1,624		40.5%
Coal		773		19.3%
Petroleum (total)		17		0.4%
Nuclear	**790**		**19.7%**	
Renewables (total)	**792**		**19.8%**	
Wind		338		8.4%
Hydropower		291		7.3%
Solar (total)		91		2.3%
Photovoltaic		88		2.2%
Solar Thermal		3		0.1%
Biomass (total)	**56**		**1.4%**	
Geothermal	**17**		**0.4%**	
Total – All Sources	**4,007**		**100%**	

Image provided by the U.S. Energy Information Administration, https://www.eia.gov/tools/faqs/faq.php?id =42 7&t=3

In 2020:

> "4.01 trillion kilowatt hours (kWh) of electricity were generated at electricity generation facilities in the U.S. and a combined total of 60.6% of the generation of electricity came from fossil fuels, coal, natural gas, petroleum, and other gases; 19.7% came from nuclear energy; the remaining 19.8% came from renewable energy sources." (US Energy Information Administration 2021).

Threats to the energy sector

The electricity component has several risks which are

> "natural disasters and extreme weather conditions; an aging workforce and human errors; equipment failure and aging infrastructure; evolving environmental, economic, and reliability regulatory requirements; and changes in the technical and operational environment, including changes in fuel supply." (US Energy Information Administration 2021).

The oil and natural gas component views many of the same threats with the addition of regulatory and legislative changes, domestic and international, due to the foreign locations of many of the oil and natural gas facilities. Those threats are

> "environmental and health; a volatile oil and gas market; operational hazards including blowouts, spills and personal injury; disruptions due to political instability, civil unrest, or terrorist activities; and transportation infrastructure constraints impacting the movement of energy resources." (US Energy Information Administration 2021).

Cybersecurity is an evolving security challenge for the energy sector, as well as for the other fifteen sectors of critical infrastructure that is equally pertinent to the global community. For the US, the private ownership necessitates a collaborative effort with the federal government and private companies to secure the facilities and networks upon whom they depend. Natural disasters and extreme weather instances demand that the physical security and resilience of energy systems and facilities must be of utmost concern. In 2012, severe windstorms in the Midwest and mid-Atlantic resulted in power outages for five million customers; "Hurricane Sandy, also in 2012, resulted in power outages for 10 million customers in 17 states; lastly, in 2013 bad actors used high powered rifles to destroy power transformers at a substation in California, which did not result in power outages for customers, but did result in more than $15 million in damages." (US Energy Information Agency 2021).

A DHS study concluded that many aspects of the US critical infrastructure are approaching the end of their projected life spans, and many are in dire need of modernizing through technological innovations and increased resilience while addressing current and future customer needs (US Energy Information Agency 2021). As the labor pool retires, replacing it with skilled labor is an area of great concern requiring long-term investment for recruiting, training, and patience.

Partner sectors

There exists a high degree of interdependence within the energy sector as well as with the other fifteen sectors of critical infrastructure. The energy sector fuels the nation and is dependent upon the nation's transportation, information technology, communications, finance, and government sectors. The US energy systems and multiple networks extend beyond the nation's borders, which requires international collaboration to secure the energy sectors infrastructure.

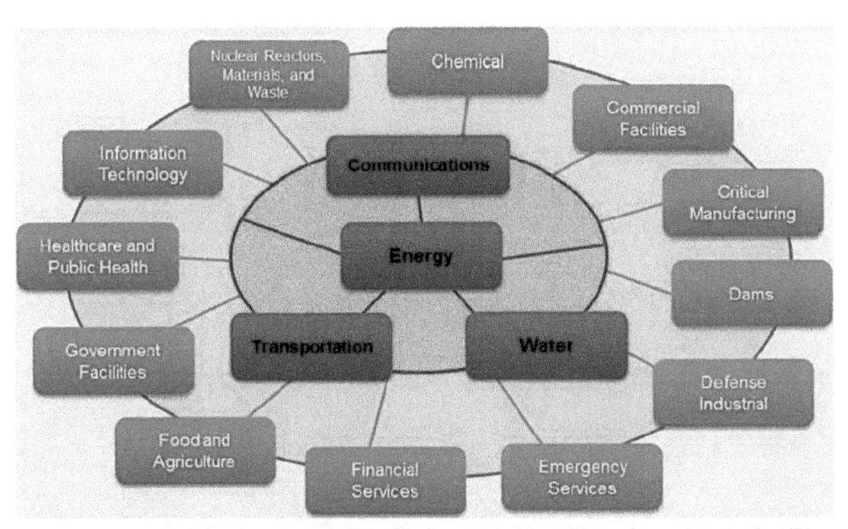

Image courtesy of CISA, Energy Sector-Specific Plan, 2015; https://www.cisa.gov/sites/default/files/publications/nipp-ssp-energy-2010-508.pdf

The energy sector depends on a collaborative effort between industry, FSLTT echelons of government, and other sectors of critical infrastructure. The energy sector provides fuel to all critical infrastructure sectors; thus, the energy sector is critical to the national security and economy that a disruption or loss of energy function will directly affect the security and resilience of other critical sectors of infrastructure. Yet the energy sector depends on many of the other sectors, such as transportation, IT, communications, water, financial services, and government facilities.

Chapter 11
Financial Services Sector

The organizations that make up the financial services sector "form the backbone of the nation's financial system and are a vital component of the global economy tied together through a network of electronic systems" (CISA 2015g). The financial services sector consists of thousands of

> "depository institutions, providers of investment products, insurance companies, other credit and financing organizations, and the providers of the

> critical financial utilities and services that support these functions." (CISA 2015g)

The US has the largest financial market in the world, and the "2021 estimated industry-wide revenue was $4.8479 trillion" (Statista 2021). The financial institutions in the US vary in size from small community banks to global powerhouses with thousands of employees and billions of dollars in assets. The services that the financial institutions provide determines the government entities that regulate them. The financial services sector encompasses multiple services in four primary categories: "deposit, consumer credit and payment systems products; credit and liquidity products; investment products; and risk transfer products" (CISA 2015g).

Deposit, consumer credit, and payment systems products

The depository institutions are the

> "primary providers of wholesale and retail payments services, such as wire transfers, checking accounts, and credit and debit cards; they facilitate the conduct of transactions across the payments infrastructure, including electronic large value transfer systems, automated clearing houses (ACH), and automated teller machines (ATM) and are the primary means of contact by citizens with the financial services sector." (CISA 2015g).

Depository institutions are also involved in the lives of many citizens because they provide "forms of credit, such as mortgages and home equity loans, collateralized and uncollateralized loans, and lines of credit, including credit cards" (CISA 2015g). Bank customers can now, thanks to technology, make deposits in person or over the Internet via computers or on mobile devices. While this is incredibly convenient, it is also risky as it is an opportunity for bad actors to hack into personal accounts.

Credit and liquidity products

Every day, customers may need to secure a loan, such as an "automobile loan or mortgage to purchase a home; businesses may need to open a line of credit to expand their operations; governments may issue sovereign debt obligations to fund operations or manage monetary and economic policy" (CISA 2015g). These credit needs will be processed by "depository institutions, finance and lending firms, securities firms, and government-sponsored enterprises; sometimes these financial entities will provide credit directly to the customer, other times the credit will be extended indirectly by financial service firms that provide these services on a retail basis, say to purchase new kitchen appliances." (CISA 2015g).

Investment products

The sheer number of investment service providers and a wide variety of products encourage the global competitiveness of US financial markets. The investment products can be short- or long-term investments; they can include "debt securities such as bonds and bond mutual funds; equities such as stocks or stock mutual funds; exchange-traded funds, and derivatives such as options and futures" (CISA 2015g).

Risk transfer products

The transfer of financial risks is critical for the economic vitality and sustainability of businesses and individuals due to factors such as theft, the destruction of physical or electronic property, cybersecurity incidents, or the loss of income due to a death or disability in a family. Many US financial institutions provide risk transfer products, making it the largest in the world, estimated to be trillions of dollars. Multiple financial products are available that allow customers to transfer diverse types of financial risks and are offered by insurance companies and futures firms. The hedging of risk occurs when market participants engage in financial investments and financial risk transfers.

Sector risks

Financial institutions face evolving risks including "operational, liquidity, credit, legal, and reputational risk" (CISA 2015g). The financial service sector organizations "form the backbone of the nation's financial system and are a vital component of the global economy" (CISA 2015g). These organizations are connected via a network of electronic systems. In fact, most of the sector's services rely on information and communications technology platforms, increasing the risk posed by cyberattacks.

Natural disasters such as hurricanes, tornadoes, and floods have the potential to cause disruptions to the physical equipment that the financial services sector relies on. Three recent incidents negatively affect the ability of the sector to conduct operations.

1. The September 11, 2001, attacks caused the securities markets and several futures exchanges to close until communications and other services were transferred to alternate sites or restored to lower Manhattan.
2. In the summer of 2012, financial institutions, including smaller institutions, experienced a series of coordinated distributed denial-of-service (DDoS) attacks against their public-facing websites. These incidents affected customer access to banking information but did not impact core systems or processes.
3. On October 29, 2012, the landfall of Superstorm Sandy caused a two-day closure of major equities exchanges, while fixed income markets were closed for one day. In recent years, cybercriminals have accessed numerous retailers and other networks to steal credit card information and other financial data. (CISA, 2015g)

According to CISA, it is essential to continually assess and understand the constant and evolving threat posed by nation-states and bad actors, possibly supported by nation-states, to our financial services sector.

> "The evolving threat of cybersecurity and physical risks to this sector to include the identification of critical processes and their dependence on information technology and supporting operations for the delivery of financial products and services. Given that financial institutions and technology service providers are tightly interconnected in a dynamic marketplace, an incident impacting one firm has the potential to have cascading impacts that quickly affect other firms or sectors. This risk is exacerbated by the fact that financial institutions depend on other sectors for key services like electricity, communications, and transportation. Under the EO 13636 Section 9 framework, owners, and operators of identified critical infrastructure whose business and operations depend on an extensive network of information and communications technology and software (or "cyber dependent") may be eligible for expedited processing of clearance through the DHS Private Sector Clearance Program, which may provide access to classified government cybersecurity threat information as appropriate." (CISA 2015g).

Critical infrastructure partners

There are many threats to the financial services sector, including cybersecurity and physical ones, but there are also sector dependencies with the energy, communications, information technology, and transportation sectors, all of whom are critical to the ability of the financial sector to function.

Chapter 12
Food and Agriculture Sector

The National Strategy for Physical Protection of Critical Infrastructures and Key Assets defines the food and agriculture sector as "the supply chains for feed, animals, and animal products, crop production and the supply chains of seed, fertilizer, and other necessary related materials; and the post-harvesting components of the food supply chain, from processing, production, and packaging through storage and distribution to retail sales, institutional food services, and restaurant or home consumption." (National Strategy for Physical Protection of Critical Infrastructures and Key Assets 2003).

The food and agricultural sector consist of production, processing, and delivery systems that feed both people and animals globally and account for nearly one-fifth of the nation's economic activity (CISA 2015h). According to CISA, "in 2012, the total of agricultural product sales amounted to $400 billion; in 2013, one-fifth of all U.S. agricultural production was exported, generating $144.1 billion in revenue" (CISA 2015h). The US food and agriculture systems "are almost entirely under private ownership, operate in highly competitive global markets, strive to operate in harmony with the environment, and provide economic opportunities and an improved quality of life for American citizens and others worldwide." (CISA 2015h)

In 2014, there were more than "935,000 restaurants and institutional food service establishments; an estimated 114,000 supermarkets, grocery stores and other food outlets; 81,575 Food and

Drug Administration (FDA) registered domestic food facilities such as warehouses, manufacturers, processors; and 115,753 FDA registered foreign food facilities." (CISA 2015h).

The sheer size and expanse of the food and agriculture sector is truly immense. The US Department of Agriculture (USDA) Food Safety and Inspection Service (FSIS) "regulate 6,756 establishments for meat, poultry, processed egg products, imported products, and voluntary inspection services; the U.S. has roughly 2.1 million farms, encompassing 915 million acres of land; U.S. farms produce $212 billion in crop production." (CISA 2015h)

The top five industries within this sector are "cattle, poultry and eggs, corn, soybeans, and milk; the majority of the sector is heavily involved in food production, but the food agriculture sector also imports many ingredients and finished products from around the globe, thus creating a complex labyrinth of growers, processors, suppliers, transporters, distributors, and consumers." (CISA 2015h).

Faced with supply and demand criteria, this sector must also factor in global logistical challenges, changes in various nation's agricultural import and export policies, and new technological developments in the cultivating and harvesting of multiple products by US farmers.

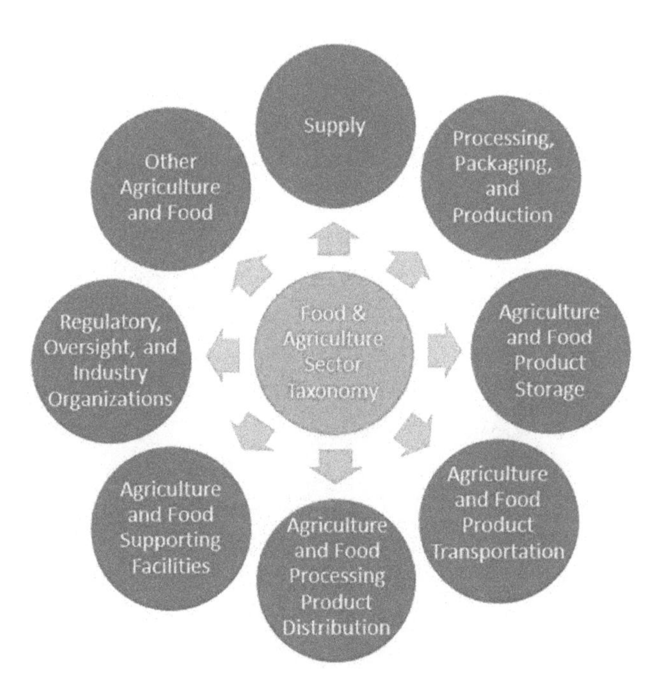

DHS developed "an infrastructure data taxonomy to enable transparent and consistent communication regarding critical infrastructure between government and private sector partners" (CISA 2015h). The taxonomy is divided into several categories, which are listed below.

The food and agriculture sector taxonomy defines agriculture and food as:

> "Agriculture comprises establishments primarily engaged in growing crops, raising animals, harvesting timber, and harvesting fish and other animals from a farm, ranch, or their natural habitats. Food establishments transform livestock and agricultural products into products for intermediate or final consumption. The industry groups are distinguished by the raw materials (generally of animal or vegetable origin) processed into food

> and beverage products. The food and beverage products manufactured in these establishments are typically sold to wholesalers or retailers for distribution to consumers." (CISA 2015h)

Interestingly, "both the USDA and FDA, a subordinate agency within the Department of Health and Human Services (HHS), share regulatory responsibility for food" (CISA, 2015h). The USDA is responsible for "the regulation of meat, poultry, and processed egg products in accordance with the Federal Meat Inspection Act (FMIA), Poultry Products Inspection Act (PPIA), and Egg Products Inspection Act (EPIA)" (CISA 2015h). The FDA has responsibility for "the remaining food products not under the regulatory authority of USDA and food is defined in Section 201(f) of the Federal Food, Drug, and Cosmetic Act (FFDCA) as (1) articles used for food or drink for man or other animals, (2) chewing gum, and (3) articles used for components of any such article." (CISA 2015h).

Both the USDA and FDA "share sector responsibilities for the safety and defense of agriculture and food and have an obligation to provide leadership for sector infrastructure security and resilience activities, which include establishing information-sharing relationships and developing collaborative sector protection plans with sector critical infrastructure partners" (CISA 2015h).

The FDA is responsible for "the safety of 80 percent of all food consumed in the U.S. and both agencies have been assigned responsibility for overseeing and coordinating security and resilience efforts" (CISA 2015h).

Threats to the food and agriculture sector

One of the biggest threats to the food and agriculture sector is that of food contamination and disruption, either accidental or intentional. It has been estimated that "contaminated food is responsible for approximately 48 million illnesses, 128,000 hospitalizations, and 3,000 deaths in the U.S., at a cost of more than $14 billion a year in terms of medical care, lost productivity, chronic health problems, and deaths." (CISA 2015h).

Another threat would be that posed by terrorists who consider this sector to be an especially tempting target. If a bad actor were to poison the food supply, it would be costly in terms of economic losses, it would create public panic, and lead to a "public health crisis with considerable mortality and morbidity" (CISA 2015h). Extreme weather and "climate characteristics, such as temperature, precipitation, carbon dioxide, and water availability, directly impact the health and wellbeing of plants and livestock, pasture and rangelands, water resources, agriculture, land resources, and biodiversity" (CISA 2015h). Cyber threats evolve rapidly and could attack the networks that manage livestock feeding, immunization, and crop fertilization and watering, thus this sector requires constant monitoring and updating of computer networks and security programs. Insider threats require that farms and food processing centers must have state-of-the-art intrusion detection and intrusion prevention systems and surveillance programs.

Partner sectors

The food and agriculture sector has dependencies and cannot function without assistance from the following sectors: energy, information technology, water and wastewater, chemical, communication, financial, transportation, and the health care and public health sectors.

Chapter 13
Government Facilities Sector

The government facilities sector does not generate product nor revenue, but it is the means by which the federal government governs the nation, and all of its facilities are uniquely governmental. This one sector is at the heart of each of the other fifteen sectors of critical infrastructure. It is also the most important in that without governance, one cannot have a nation-state capable of functioning and interacting with others on the global stage. Government facilities exist to facilitate citizens to conduct business with their government, even if it is as mundane as getting a driver's license or to apply for benefits. The government facilities sector includes more than "900,000 constructed assets owned or operated by the Federal Government alone" (CISA 2015i). Interestingly, the assets owned or operated by the fifty States and six territories "encompass 3,031 counties, 85,973 local governments, and 566 federally recognized tribal nations" (CISA, 2015i) also fall in the purview of the government facilities sector This is one of the "largest and most complex sectors within the National Infrastructure Protection Plan 2013 (NIPP) framework and includes a variety of facilities owned or leased by the FSLTT governments located both within the U.S. and overseas" (CISA 2015i).

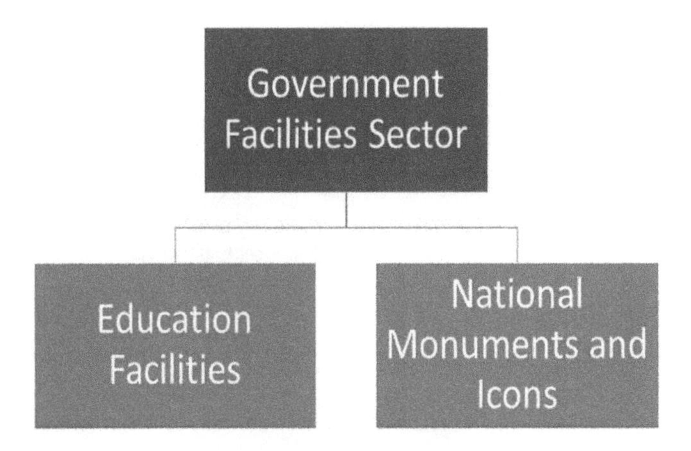

As previously stated in chapter 2, with the issuance of Presidential Policy Directive 21 (PPD-21), critical infrastructure was designated, and its importance was explained. The PPD initiated a national effort to better secure the sixteen sectors of critical infrastructure. PPD-21 also designated "DHS and the GSA to serve as co-leads for the government facilities sector and as the federal points of contact for the sector" (CISA 2015i). The designation of co-leads for the sector was based on the expertise that each agency possessed. PPD-21 also directed that the creation of two subsectors: "the Education Facilities Subsector, which covers schools, institutions of higher education, and trade schools; and the National Monuments and Icons Sector be included in the government facilities sector" (CISA 2015i). The government facilities sector is "focused on the threats and hazards that are likely to cause harm and employs prioritized approaches that are designed to prevent or mitigate the effects of incidents, which allows operators to make risk-informed decisions that best allocate limited resources and increases security and strengthens resilience by identifying and prioritizing actions to ensure continuity of essential functions and services during incidents and to support rapid response and restoration." (CISA 2015i).

Public Facilities

- Offices and office building complexes
- Housing for government employees
- Correctional facilities
- Embassies, consulates, and border facilities
- Education facilities
- Courthouses
- Maintenance and repair shops
- Libraries and archives
- Monuments

Non-Public Facilities

- Research and development facilities
- Military installations
- Record centers
- Space exploration facilities
- Storage facilities for weapons and ammunition, precious metals, currency, and special nuclear materials and waste
- Warehouses used to store property and equipment

There are some government facilities that are exclusive to the government facilities sector, such as headquarters buildings belonging to the federal government's various departments and agencies, but there are plenty of government facilities in the other fifteen sectors of critical infrastructure. A great many of the government facilities are open to the public for a variety of purposes such as "business activities, commercial transactions, provision of services, or recreational activities" (CISA 2015i). Yet there are also other government facilities that are not open to the public because they "contain extremely sensitive information, materials, processes, and equipment with classified missions" (CISA 2015i). The personnel working there require extensive background checks, a better-than-average lifestyle with a good credit report, minimum experience as a defendant with law enforcement and the courts system, and a federally investigated and authorized security clearance to gain employment and access.

The image below neatly demonstrates the dependencies and interdependences of the government facilities sector with other sectors of critical infrastructure. The energy sector provides power for the

Predominant Use		Responsible Sector
	Offices and office building complexes	Commercial Facilities
	Retail stores within government facilities, government agencies within commercial facilities	
	Housing or community service facilities provided for public use	
	Food service establishments within government facilities	Agriculture and Food
	Health clinics and medical units within government facilities	Healthcare and Public Health
	Transportation-related government facilities*	Transportation Systems
	Nuclear reactors, materials, and waste located in government facilities**	Nuclear Reactors, Materials, and Waste
	Police, fire, and emergency services stations	Emergency Services
	Emergency operations, command, dispatch, and control centers	
	Public works facilities associated with:	
	Water or wastewater treatment	Water
	Power or natural gas Highway or road service or maintenance	Energy
	Telephone or Internet service	Communications, Information Technology
	Highway or road service or maintenance	Transportation Systems

** Except for space exploration and any that are part of military installations.*

*** Except for all U.S. Department of Energy (DOE) facilities involved with storage or use of special nuclear material and all U.S. Department of Defense (DoD) nuclear facilities and materials associated with defense programs.*

Image provided courtesy of the Government Facilities Sector-Specific Plan, An Annex to the NIPP 2013, dated 2015;https://www. cisa.gov/sites/default/files/publications/nipp-ssp-government-facilities-2015-508.pdf

government facilities sector. From lighting to computers to HVAC systems, energy is required for the government facility to minimally function. The water sector provides a steady supply of potable water, processes wastewater, and provides water for fire suppression systems within all government facilities. The government services sector will coordinate with the emergency services sector in the aftermath of a natural disaster or terrorist attack. Fortunately, the multiple jurisdictions that responded to the 9/11 attack upon the Pentagon had just practiced what they would do, where they would set up, how they would coordinate with one another two weeks prior to the actual attack. Thus there was no learning curve, and countless lives were saved as well as the building itself due to their quick response and well-coordinated, rehearsed actions. The communications sector provides telecommunications which are necessary for operations across the federal government and most of the world. Without it, the government facilities sector and the other fifteen sectors would be unable to function, and the cascading effect would have a catastrophic effect upon the economy and on national security.

The transportation systems sector provides for the transportation of goods, employees, and visitors to and from government facilities. After a natural or man-made disaster, the transportation systems sector must be reconstituted to get services up and running and facilities operational once again. The information technology sector enables everyday operations and financial transactions, critical to a fully functioning government facilities sector. The government facilities sector supports the health care and public health sector in the aftermath of a natural or man-made disaster or pandemic, such as the one we are living through now. The financial services sector provides essential services to the government facilities sector so it can conduct business operations and emergency responses. The commercial facilities sector resides within many government facilities, which promotes collaboration for a multitude of challenges, but their close proximity to one another also creates a mutual risk to both sectors.

Threats to the government facilities sector

Natural and man-made events are the two main threats to this sector. Natural threats such as hurricanes, earthquakes, wildfires, and tornadoes are present in somewhat predictable and well-defined geographical areas. The "ejection of material from our sun, referred to as coronal mass ejection space events which in layman terms is the expulsions of plasma and magnetic field from the Sun's corona, can have detrimental effects on Space Weather and Global Positioning Systems, Electric Power Transmission, High Frequency Radio Communications, and Satellite Communications" and other satellites such as those belonging to the government's GPS system." (National Oceanic and Atmospheric Administration, n.d.)

Man-made threats can be caused by human errors or carelessness, criminal or terrorist acts, active shooter, cyberattack, denial of services, insider threats, or UAS incursions. Cyberattacks can focus on the building and access systems, such as elevators, electric power, HVAC systems. These systems are usually not equipped with state-of-the-art security systems, thus by hacking into them, a bad actor can leapfrog into government networks that have some degree of connectivity with the building systems. Aging infrastructure will only increase the vulnerabilities of the federal facilities sector to natural and man-made disasters. Pandemics could affect the workforce, especially when employees congregate in offices and buildings. A pandemic could also result in psychosomatic symptoms in employees, resulting in increased absenteeism and work outages due to anxiety with the workforce.

Chapter 14
Health Care and Public Health Sector

The health care and public health sector provides "goods and services integral to maintaining local, national, and global health security with five core mission areas: prevention; protection; mitigation; response; and recovery" (CISA 2015j). The health care and public health sector infrastructure is "dedicated to building and sustaining community health resilience; enhancing and expanding the Nation's medical capacity for everyday healthcare; improving health related situational awareness capabilities; enhancing the integration of healthcare and public health capabilities into emergency management systems in effective ways; and strengthening global health security." (CISA 2015j).

The health care and public health sector spans is critical in maintaining national health security and encompasses "both the public and private sectors; it includes publicly accessible healthcare facilities, research centers, suppliers, manufacturers, and other physical assets and vast, complex public-private information technology systems required for care delivery and to support the rapid, secure transmission and storage of large amounts of healthcare and public health data." (CISA 2015j).

Managing the hazards and risks to critical infrastructure in the health care and public health sector requires a comprehensive and integrated approach.

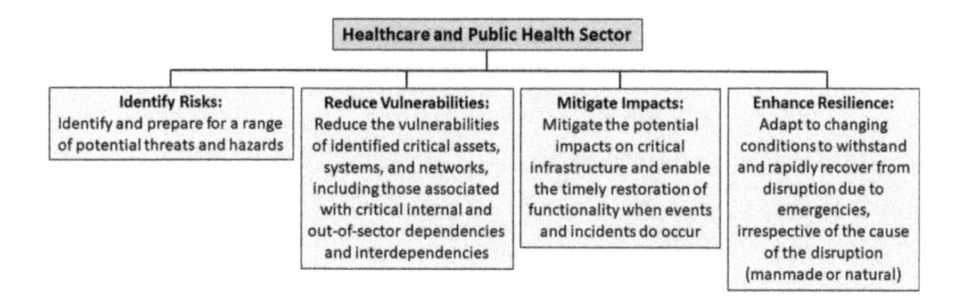

More than "10 percent of the U.S. workforce, or 14 million people," (Bureau of Labor Statistics 2014) are employed in the health care and public health sector, which includes service providers and those who play a supporting role, such as vaccine manufacturers.

The importance of this sector to the public are staggering:

> "in 2011, Americans made 262 million visits to hospital emergency or outpatient departments; nearly 50 percent of Americans require one or more prescription medications to mitigate health issues; in 2012, America's 15,673 certified nursing homes operated at over 80 percent capacity, and, at any one time, over 60 percent of the beds in America's 4,973 community healthcare facilities were occupied." (CISA 2015j).

Due to the demand for health care and public health sector facilities and providers, any disruptions to this sector's ability to meet the needs of the public could result in a detrimental impact that would affect other critical infrastructure sectors across the nation. The health care and public health sector consists of six private subsectors and two government ones.

Private subsectors: direct patient care

This is the "largest of the six subsectors and includes healthcare systems, professional associations, and a wide variety of medical facilities, public health, and emergency medical services" (CISA, 2015j). It employs over "12 million people and supports 5,686 registered healthcare facilities with more than 900,000 staffed beds; over 35 million citizens are admitted to these facilities annually" (CISA 2015j).

Health information technology

This subsector includes "medical research institutions, information standards bodies, and electronic medical record systems vendors; nearly 59 percent of America's hospitals, 95 percent of America's community pharmacies, and 40 percent of America's office-based physicians have adopted electronic health records." (CISA 2015j)

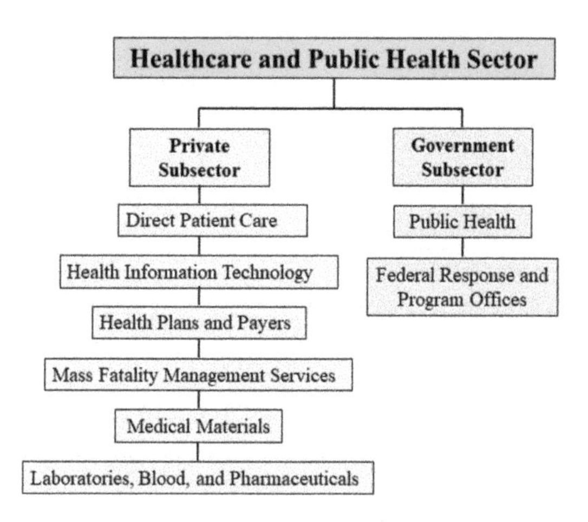

Health plans and payers health

In the US, "insurance companies and plans, local and state health departments and emergency health organizations employ over 500,000 Americans. Medicare, Medicaid, and Children's Health Insurance programs cover more than 100 million Americans" (CISA 2015j).

Mass fatality management services

Nearly "133,000 Americans work in cemetery, cremation, morgue, and funeral home occupations; includes coroners, medical examiners, forensic examiners, and psychological support personnel; approximately 86 percent of funeral homes are owned by families, individuals, or small companies" (CISA 2015j).

Medical materials

[The] "medical supply chain depends upon the 600,000 Americans who work in the public and private sectors in the areas of medical equipment and supply manufacturing and distribution; pharmaceutical distributors alone deliver 15 million prescription medicines and healthcare products to more than 200,000 licensed healthcare providers in all 50 states." (CISA 2015j)

Laboratories, blood, and pharmaceuticals

This subsector consists of "government and private assets and is critical for healthcare situational awareness, including pharmaceutical manufacturers, drug store chains, pharmacists' associations, public and private laboratory associations, and blood banks; nearly 95 percent of the pharmacies are actively prescribing and over 32 percent of new prescriptions are sent electronically." (Bureau of Labor Statistics 2014)

Government subsectors: public health

It is a collaborative effort across the FSLTT public health programs whose focus is "to improve the health of populations through education, policy, and community services; governmental public health services provide epidemiological surveillance, preparedness planning, emergency response, laboratory testing and coordination, health information communication and outreach, and programs that build community resilience." (CISA 2015j)

Public health networks provide "local hazard and risk assessments, develop mitigation plans and strategies, facilitate joint public-private sector planning and exercising, and conduct response and recovery operations" (CISA 2015j).

Federal response and program offices

The federal critical infrastructure protection partnership "relies on policy development, funding opportunities, and coordinating activities via the Government Coordinating Council (GCC), that includes diverse federal partners from the interagency HHS, DoD, and other sectors working to improve resilience of the system and support ongoing healthcare and public health operations." (CISA 2015j)

Threat to health care and public health sector

During pandemics and other health crises, "emerging diseases and their mutations can put the healthcare and public health sector personnel at risk and severely impact ongoing healthcare and public health operations" (CISA 2015j). The US is constantly assessing and attempting to respond quickly to emerging health threats such as the "COVID-19 pandemic, the severe acute respiratory syndrome (SARS), and middle east respiratory syndrome (MERS)" while battling ongoing problems such as antibiotic resistance and supply shortages (CISA 2015j). One of the biggest health threats is the antibiotic-resistant bacteria that "threaten patients, healthcare workers, suppliers of diagnostic equipment, and health facilities" (CISA 2015j). According to "The White House National Action Plan for Combating Antibiotic-Resistant Bacteria, the estimates are that at least two million illnesses and 23,000 deaths are caused by drug-resistant bacteria in the U.S. each year" (CISA 2015j).

The health care and public health sector has representatives at every echelon of health-care communities across the nation. The sector's facilities, employees, information systems, and supply chains are vulnerable to geographical natural disasters and extreme weather.

The malicious dissemination of biological or chemical agents, use of a radiological, nuclear, or explosive device, by extremist groups or international terrorist organizations would likely inflict mass casualties while attracting media attention and leading to local, regional, or national disruption of services. In addition, lone actors or active shooter threats pose a persistent problem to this sector.

Decades past the use of just-in-time supply chains became the norm. This is a threat in times of crisis to the health care and public health sector which operates with limited inventories, thereby making health-care facilities and providers especially sensitive to the cascading effects of late deliveries due to region disasters or other types of outages. Private sector entities make up "nearly 92 percent of the healthcare and public health sector and are especially vulnerable if supplies are unable to reach private sector healthcare providers and facilities, or if reach-back support is negatively impacted." (CISA 2015j)

The use of worldwide rapid transportation networks has had the unintentional consequence of "rapidly disseminating diseases, adulterated pharmaceutical supplies, tainted blood products, or contaminated food widely with unprecedented speed" (CISA 2015j).

The health care and public health sector is dependent upon information technology, both to store and transmit personally identifiable health information, sharing medical information to aid in patient care, maintaining patient records, and to conduct financial operations. Recently, nearly half of all US pharmaceutical and life science organizations experienced a cyberattack in which bad actors may have aimed to steal personal data, corrupt information, or impact financial security. Industry experts have identified malware, sophisticated viruses, intellectual property theft, and advanced persistent threats as the most significant security threats to the pharmaceutical industry and to the health care and public health sector.

There is an ever-present threat by the effects of space weather, specifically electromagnetic pulse (EMP) risks originating from our sun or even from man-made sources. Various technologies on our planet are vulnerable, such as the nation's power grid.

> "[A] sudden burst of EMP radiation is produced as part of the normal cyclical activity of the magnetic storms that flare from the surface of the sun and depending on the severity and impacted area, an EMP event could result in catastrophic repercussions for healthcare facilities, should long-term power outages" last far longer than the ability of the sector's backup power sources could function." (CISA 2015j)

Partner sectors

The entire sixteen sectors of critical infrastructure require energy to operate, and the health care and public health sector is no exception. Energy is required by all sectors of critical infrastructure, especially those that are closely tied to a functioning and capable health care and public health sector, such as the water, communications, and transportation sectors, which provide "much-needed supplies, raw materials, pharmaceuticals, personnel, emergency response units, and patients" (CISA 2015j). Water is a basic staple for all life and is vital to human health. This sector relies on

> "potable water and wastewater for infection control, sanitation, renal dialysis, laboratory needs, heating, and air conditioning, manufacturing and storage of pharmaceuticals, sterilization, maintenance of blood and organ banks, drinking water for staff, and a myriad of other uses." (CISA 2015j)

The health care and public health sector require "communications infrastructure to for situational awareness and to coordinate healthcare activities during steady state and emergencies" (CISA 2015j). The emergency services sector consists of "emergency services facilities and associated systems, as well as trained and tested personnel to provide life safety and security via the first-responder community" (CISA 2015j). It is closely intertwined with the health care and public health sector in that it supports emergency response preparedness and operations. The health care and public health sector depends on the "emergency services sector as the nation's first line of defense and prevention, and for its role in the short-term mitigation of consequences immediately following a disaster" (CISA 2015j). Speed and coordination with the emergency services sector in the aftermath of a disaster is critical to the success of the health care and public health sector's life-saving efforts.

Chapter 15
Information Technology Sector

The information technology sector provides "products and services that support our global information-based society" and are critical to the other fifteen sectors of critical infrastructure" (CISA 2016). In fact, information technology sector provides "the information technology infrastructure upon which all other critical infrastructure sectors rely, as such the information technology sector's vision is to achieve a sustained reduction in the impact of incidents on the sector's critical functions" (CISA 2016). The information technology sector is comprised of small, medium, and large companies, some of which are multinational. The information technology sector is "functions-based, comprised of physical assets, networks, and other virtual systems that enable key capabilities and services in both the public and private sectors" (CISA 2016).

There are six critical functions of the information technology sector's that provide IT products and services for the other fifteen sectors of critical infrastructure, which are required to "maintain or reconstitute networks such as the Internet, local networks, wide area networks, and associated services" (CISA 2016). These functions are vital to the nation's economic security and public health, safety, and confidence and are distributed across a broad network of infrastructure designed to withstand and rapidly recover from most threats. These critical information technology sector functions are provided by "a combination of owners and operators who provide information technology hardware, software, systems, and services to include the development, integration, operations, communications, testing, and security" (CISA 2016).

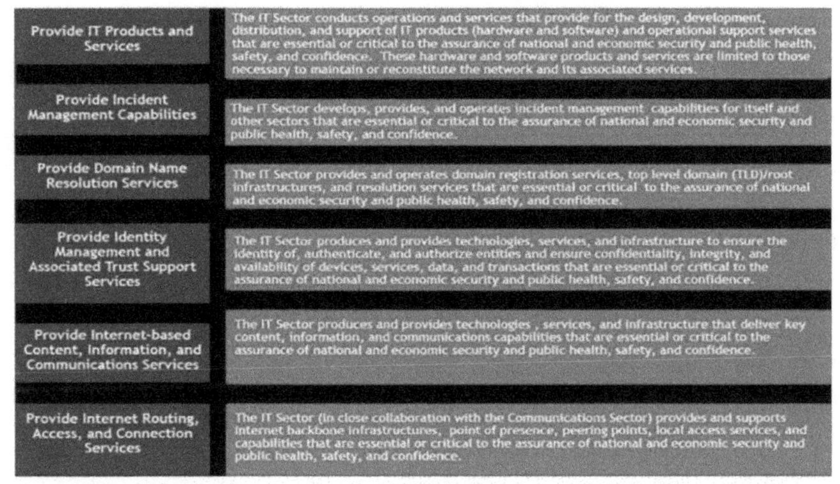

Provide IT Products and Services	The IT Sector conducts operations and services that provide for the design, development, distribution, and support of IT products (hardware and software) and operational support services that are essential or critical to the assurance of national and economic security and public health, safety, and confidence. These hardware and software products and services are limited to those necessary to maintain or reconstitute the network and its associated services.
Provide Incident Management Capabilities	The IT Sector develops, provides, and operates incident management capabilities for itself and other sectors that are essential or critical to the assurance of national and economic security and public health, safety, and confidence.
Provide Domain Name Resolution Services	The IT Sector provides and operates domain registration services, top level domain (TLD)/root infrastructures, and resolution services that are essential or critical to the assurance of national and economic security and public health, safety, and confidence.
Provide Identity Management and Associated Trust Support Services	The IT Sector produces and provides technologies, services, and infrastructure to ensure the identity of, authenticate, and authorize entities and ensure confidentiality, integrity, and availability of devices, services, data, and transactions that are essential or critical to the assurance of national and economic security and public health, safety, and confidence.
Provide Internet-based Content, Information, and Communications Services	The IT Sector produces and provides technologies , services, and infrastructure that deliver key content, information, and communications capabilities that are essential or critical to the assurance of national and economic security and public health, safety, and confidence.
Provide Internet Routing, Access, and Connection Services	The IT Sector (in close collaboration with the Communications Sector) provides and supports internet backbone infrastructures, point of presence, peering points, local access services, and capabilities that are essential or critical to the assurance of national and economic security and public health, safety, and confidence.

Image courtesy of Information Technology Sector-Specific Plan: An Annex to the NIPP 2013; https://www.cisa.gov/sites/default/files/publications/nipp-ssp-information-technology-2016-508.pdf

Over the past five years, the information technology sector has changed thanks to technological advancements such as the

> "increased adoption of cloud computing by enterprises and consumers; massive growth in mobile computing and mobile applications for smartphones and tablets; expanding awareness and deployment of the Internet of Things (IoT) and the trend in smart sensors and smart devices controlling physical systems in the cyber physical domain; extensive organizational acceptance of Bring Your Own Device (BYOD) in the corporate environment and the collapse of the perimeter as a defense; increasing reliance on advanced analytics; ever-increasing information technology operational complexity; unrelenting, global rise in the demand for information technology services and products; other trends include the

> widespread use of server virtualization in data centers, massive growth in social networking, continuing emergence of software defined networking, and the continuing arrival of wearable technology and 3-D printing." (CISA 2016)

Partner sectors

The information technology sector relies on the energy sector for its steady and constant supply of energy. It is also reliant on the defense industrial base which drives a great of information technology research and development, the transportation system to move information technology goods across the country and globally, and the financial services sector to handle the financial transactions that in large part fund the information technology sector.

Threats to the information technology sector

All fifteen sectors of US critical infrastructure are dependent upon globally connected information technology networks, architecture, and systems to operate. This capability allows the sectors to flourish, and their success feeds the nation's economy, the safety of its citizens, national security, and its resilience to attack. This is equally true in the global community. The IoT was mentioned in a "National Security Telecommunications Advisory Committee report that stated that the IoT will increase in both speed and scope and will impact virtually all sectors of our society" (CISA 2016). The threats to the information technology sector are evolving, are complex, and represent a serious threat to national security. Cyber and physical dangers due to criminals, hackers, terrorists, and nation-states are the biggest threats to the information technology sector. Many of the threats have been demonstrated in areas around the globe. Cyber threats have evolved into extraordinarily complex attacks that exploit known and unknown vulnerabilities in information technology products and services.

Chapter 16

Nuclear Reactors, Materials, and Waste Sector

The US nuclear sector consists of

> "a wide variety of assets, systems, and networks; in 2020, there were 94 operating commercial nuclear reactors at 56 nuclear power plants in 28 states whose cumulative efforts produced 96,555 MW of electricity, or 19.7 percent of the nation's electrical requirements." (US Energy Information Administration 2020)

State	Number of Operating Reactors	Total Nuclear Power Generated in Thousand Megawatt Hours	Percentage of State Requirement for Electricity Produced by Nuclear Power
Alabama	5	42,651	42%
Arizona	3	32,340	31%
Arkansas	2	12,691	21%
California	4	17,901	9%
Connecticut	2	16,499	48%
Florida	5	29,146	12%
Georgia	4	33,708	26%
Illinois	6	97,191	53%
Iowa	1	5,213	9%
Kansas	1	10,647	21%
Louisiana	2	15,409	16%
Maryland	2	15,106	44%
Massachusetts	1	5,047	16%
Michigan	3	32,381	29%
Minnesota	3	13,904	24%
Mississippi	1	7,364	12%
Missouri	1	8,304	10%
Nebraska	2	6,912	20%
New Hampshire	1	9,990	57%
New Jersey	4	34,032	45%
New York	6	42,167	33%
North Carolina	7	42,374	33%
Ohio	3	17,687	15%

Note: the data was compiled from the US Energy Information Administration, 2019.

France has fifty-seven nuclear power plants that produce 75% of its annual electrical production requirements, by far the highest for any country (Ranson 2017). Interesting facts about the global nuclear energy production by country:

1. In 2020, nuclear power produced 4.3% of the global energy requirements amounting to 2.553 TW of energy produced.
2. The following countries produced the stated percentage of the global nuclear energy production:
   a) U.S.—30.9%
   b) China—13.5%
   c) France—13.3%

   d) Russia—7.9%
   e) Republic of Korea—6%
   f) Canada—3.6%
   g) Ukraine—2.8%
   h) Germany—2.4%
   i) Spain—2.2%
   j) Sweden—1.9%
   k) U.K.—1.8%
   l) Japan—1.7%
   m) India—1.6%
   n) Belgium—1.3%" (Bhutada 2022)
3. China is planning to build 150 new reactors in the next 15 years at an estimated cost of $440 billion. (Venditti 2021)

The nuclear reactors, material, and waste sector is well known for the nuclear power plants that produce electricity, but it also consists of the

> "creation and processing of fuel used in nuclear power plants; radioactive materials used tens of thousands of times every day for a wide variety of medical, research; industrial purposes, and radioactive waste and disused radioactive sources; and the decommissioned reactors and the protection of nuclear materials still in storage at these sites." (CISA 2015k)

These materials are "created, or imported, distributed, and transported throughout the nation, and they are important components of the nation's Chemical, critical manufacturing, energy, food and agriculture, and healthcare and public health sectors" (CISA 2015k).

Nuclear Power Plants:
- Boiling water reactors (BWR); and
- Pressurized water reactors (PWR).

Research, Training, and Test Reactors:
- Government research and test reactors;
- University research and training reactors; and
- Private research and test reactors.

Decommissioned Nuclear Facilities:
- Deactivated reactors; and
- Other deactivated nuclear facilities.

Fuel Cycle Facilities:
- Uranium mining or in situ uranium leaching;
- Uranium ore milling or leachate processing;
- Uranium conversion facilities;
- Uranium enrichment facilities;
- Fuel fabrication facilities:
  - Category I (special nuclear materials) facilities;
  - Category II (special nuclear materials—moderate strategic significance) facilities; and
  - Category III (special nuclear materials—low strategic significance) facilities.

Nuclear Materials Transport:
- Low-hazard radioactive materials transport; and
- High-hazard radioactive materials transport.

Radioactive Materials:
- Medical facilities with radioactive materials;
- Research facilities using radioactive materials;
- Irradiation facilities; and
- Industrial facilities with nuclear materials.

Radioactive Source Production and Distribution Facilities:
- Radioactive device manufacturers;
- Radioactive source producers;
- Radioactive source importers; and
- Radioactive source manufacturers.

Nuclear Waste:
- Low-level radioactive waste processing and storage facilities;
- Sites managing accumulations of naturally occurring radioactive materials (NORM);
- Spent nuclear fuel processing and storage facilities:
  - Spent nuclear fuel wet storage facilities; and
  - Spent nuclear fuel dry storage facilities;
- Transuranic waste processing and storage facilities;
- High-level radioactive waste storage and disposal facilities; and
- Mixed waste processing.

Image courtesy of The Nuclear Sector Taxonomy from the Nuclear Reactors, Materials, and Waste Sector-Specific Plan: An Annex to the National Infrastructure Protection Plan 2010; https://www.dhs.gov/xlibrary/assets/nipp-ssp-nuclear-2010.pdf

Research and training reactors

There are thirty-two nuclear research and test reactors in twenty-three states, most at major universities that are low-power producers yet are valuable for conducting research, developing theoretical practices, producing radioactive sources, and for educational, industrial, medical, and the training of future nuclear engineers. Research and training reactors range in size "from 0.1 watt to 20 megawatts (MW); a typical commercial nuclear power reactor is rated at 3,000 MW" (CISA 2015k). Several research and training reactors are collocated with quantities of radiological materials or utilize highly enriched uranium (HEU) fuel, both of which represent a significant danger if stolen and dispersed near populated areas and other soft targets.

The NRC has licensed

> "eight major fuel fabrication and production facilities to operate in six states; it regulates two gaseous diffusion fuel enrichment facilities, one operational and one in cold shutdown; it also regulates nine additional facilities other than reactors that possess small quantities of SNM or process source material." (CISA 2015k)

Uranium mining and milling facilities are also part of the fuel cycle facility category and consist of the following:

> "A chemical plant designed to extract uranium from mined ore. The mined ore is brought to the milling facility by truck where the ore is crushed and leached. In most cases, sulfuric acid is used as the leaching agent, but alkaline leaching can also be used. The leaching agent extracts not only uranium from the ore, but also several other constituents such as molybdenum, vanadium, selenium, iron, lead, and arsenic. The uranium product from the mill is referred to as "yellowcake" because of its yellowish color. The yellowcake generated by both processes is sent to a conversion facility for the next step in manufacturing nuclear fuel. Conventional mills crush the pieces of ore and extract 90 to 95 percent of the uranium. Mills are typically located in areas of low

> population density, and they process ore from mines within about 30 miles. In situ leach facilities are another means of extracting uranium from underground mines; they are able to recover uranium from ores that may not be economically recoverable by other methods. In this process, a leaching agent such as oxygen with sodium carbonate is injected through wells into the ore body to dissolve the uranium. The leach solution is pumped from the formation, and ion exchange separates the uranium from the solution. Twelve such facilities exist in the U.S." (CISA 2015k)

Low-level radioactive waste disposal

Low-level waste often includes items that have "become contaminated with radioactive material, have become radioactive, or have been exposed to neutron radiation" (CISA 2015k). This waste typically consists of "contaminated protective shoe covers and clothing, wiping rags, mops, filters and resins, reactor water treatment residues, equipment and tools, luminous dials, catheters, swabs, injection needles, syringes, and disused radioactive sealed sources" (CISA 2015k). The level of radioactivity can be as low as that of normal background levels found in nature to exceedingly elevated levels. It is normal and entirely legal for much of today's

> "low-level waste to be stored onsite by licensees until the radioactivity has decayed, the items can be disposed of as ordinary trash, or the waste can be accumulated and shipped to a low-level waste disposal site in containers approved by DOT or the NRC. [Commercial] low-level waste disposal facilities must be licensed in accordance with health and safety requirements; they must be designed, constructed, and operated to meet safety and environmental protection standards;

> and the operator of a facility must also extensively characterize the site where the facility is located and analyze how the facility will perform for at least 500 years." (US Nuclear Regulatory Commission 2019)

"Low-level waste disposal facilities can provide storage for chemical waste as well and when this occurs, the facilities are considered to be components of both the nuclear and chemical sectors" (US Nuclear Regulatory Commission 2019).

High-level radioactive waste disposal

High-level radioactive wastes are "the highly radioactive material produced as a byproduct of the reactions that occur inside nuclear reactors" (US Nuclear Regulatory Commission 2019). High-level wastes take one of three forms: "irradiated nuclear fuel; highly radioactive material remaining after spent fuel is reprocessed; or other highly radioactive material that the NRC, consistent with existing law, determines by rule requires permanent isolation" (US Nuclear Regulatory Commission 2019). Currently, there is no technology to render radioactive waste radioactively harmless. The only existing process is to isolate it and allow for the passage of time. Depending on the radiological materials involved and their radiological levels, this can be a brief period of time, or it can take thousands of years. Nuclear waste is radioactive and is a byproduct of "being used in nuclear reactors, fuel processing plants, hospitals, and research facilities, and while decommissioning and dismantling nuclear reactors and other nuclear facilities (US Nuclear Regulatory Commission 2019). High-level radioactive waste is usually "uranium fuel from a nuclear power reactor and is thermally hot as well as highly radioactive, thus it requires remote handling and shielding" (CISA 2015k).

Nuclear reactor fuel "contains ceramic pellets of uranium-235 inside of metal rods and these heavier-than-uranium elements do not produce nearly the amount of heat or penetrating radiation that fission products do, but they take much longer to decay" (US Nuclear

Regulatory Commission 2019). Radioactive isotopes "eventually decay into harmless materials; some isotopes decay in hours or even minutes, but others do not; Strontium-90 and cesium-137 have half-lives of 30 years, which means that half of their radioactivity will decay in 30 years; Plutonium-239 has a half-life of 24,000 years" (US Nuclear Regulatory Commission 2019).

Spent fuel storage

"All operating nuclear power reactors store spent fuel in spent fuel pools located within protected areas" (CISA 2015k). The US has

> "42 licensed independent spent fuel storage installations; 28 are with active electricity producing nuclear power plants; 8 decommissioned power plant sites; four power plant sites that are in the decommissioning process; an interim storage facility operated by DOE at the Idaho National Laboratory and the General Electric Morris facility in Illinois." (CISA 2015k)

In 1990, the "NRC amended its regulations to authorize storage of spent fuel at reactor sites in dry cask storage systems, which is almost completely passive, is simpler and uses fewer support systems than spent fuel pools" (CISA 2015k).

Radioactive sealed sources

Radioactive materials have numerous applications, such as

> "providing critical capabilities in the public health, oil and gas, electrical power, construction, and food industries; they are used to treat millions of patients each year in diagnostic and therapeutic medical procedures; they are used in many areas for technology research and develop-

> ment at academic, government, and private institutions; they are also useful in law enforcement and military applications." (CISA 2015k)

Industrial uses involve encapsulated radioactive material, and the amount of radioactive material authorized for use varies from "one one-millionth of a curie to millions of curies" (CISA 2015k). Radioactive material can be shipped in the US by rail, air, sea, and roadways "per the hazardous material transportation safety and security regulations of DOT and the NRC" (CISA 2015k). The specific responsibilities of DOT are the following:

> "It regulates shippers and carriers of hazardous material, including radioactive material; DOT is responsible for such safety requirements as vehicle safety, routing, shipping papers, hazard communications, certain packaging requirements, emergency response information, and shipper/carrier training requirements, as well as security requirements for highway route controlled quantities of radioactive material and other quantities of radioactive material that require placards for transport." (CISA 2015k)

The NRC responsibilities include the following:

> "It regulates users of radioactive material and approves design, fabrication, use, and maintenance of transportation packages for domestic radioactive material shipments; it regulates the physical protection of commercial spent fuel and copious quantities of radioactive material in transit against sabotage or other malicious acts." (CISA 2015k)

Threats

The NRC purports that nuclear power plants are exceptionally defended and physically hardened, which have been "designed to withstand such extreme events as hurricanes, tornadoes and tornado-generated missiles, and earthquakes" (CISA 2015k). In *Unmanned Systems: Savior or Threat*, I discussed the physically hardened nature of nuclear reactor buildings at our ninety-four nuclear power plants. I believe that the test that the NRC continually points to as proof positive that the reactor building is safe, even from airplane diving toward it in a fashion similar to the Twin Towers on 9/11, was flawed and that the NRC is drawing conclusions from it that the test and data simply do not support. An excerpt from the book follows:

> "In 1988, the NRC and the Muto Institute of Structural Mechanics Incorporated of Tokyo, Japan contracted Sandia National Laboratories to crash a plane into a steel-reinforced concrete wall, replicating nuclear reactors' walls at all nuclear power plants (Tulsa World, 1989). The test was designed to demonstrate that a manmade attack upon a nuclear reactor with an airliner would not breach the reactor's steel-reinforced concrete walls. This test occurred years before the terrorist attacks on September 11, 2001. A surplus U.S. Air Force F-4 Phantom fighter jet was procured, strapped to a rocket-powered sled on a rail system, and crashed into the 10-foot-thick concrete wall at 500 miles per hour (Baker, 2017). The impact produced 2.4 inches of penetration by the jet's engines into the concrete block and moved it backward four feet (Tulsa World, 1989). The jet weighed 41,500 pounds when it hit the wall, which included more than 1,200 hundred gallons of water instead of jet fuel to simulate the weight of fuel and provide proper

> mass distribution (Baker, 2017). There were two critical issues with the demonstration that cast doubt on nuclear power plants' structural integrity when attacked by hijacked jetliners. First, when the two jetliners impacted the Twin Towers and brought them crashing down, it was not the jetliners' impact that damaged the massive steel beams that ran vertically up the entire length of the buildings. Instead, it was the resulting 2,000 degrees Fahrenheit fire that caused them to weaken and collapse. This Phantom test omitted jet fuel, so there was no resulting fire that may have weakened the steel-reinforced concrete block. Second, the attack angle was not what a jetliner would fly at if it were to dive toward a fixed nuclear power plant. The fact that a horizontal small fighter jet flying at 500 miles per hour moved the 10-foot thick, steel-reinforced concrete block back four feet ought to be an area of concern. A much larger and heavier jetliner flying at an angle closer to a near-vertical dive would not permit the reactor's concrete walls to move back four or more feet due to the impact. This lack of impact absorption would result in be a cracking or breach of the walls." (Dorn 2020).

If we discount the threat to the reactor buildings, the logical soft targets for bad actors would be to the highly vulnerable power lines and power transmission stations that flow out of nuclear power plants. These are not secured, and if they were attacked, the transmission of electricity from the nuclear power plants to the metropolitan areas that it supports would cease immediately. The impact of the loss of a single nuclear power plant's electrical generation may have only a minor impact on the nation; it will probably affect the region in which it is located. A terrorist attack would be a significant security event, especially if it gained media coverage and resulted in a loss of electricity production or, worse yet, the release of radioactive material to the environment. According to the NRC, there had been fifty-seven UAS incursions over twenty-four US nuclear power plants in the past five years (Gardiner 2016; Rogoway and Trevithick 2020; Hambling, 2020).

Partners

The nuclear sector is dependent upon multiple other sectors of US critical infrastructure such as transportation, energy, emergency services, human and public health, information technology, chemical, and communications sectors. The transportation systems sector transports nuclear materials on land, water, and air. A disruption within the transportation systems sector could hinder the movement of materials in and out of nuclear facilities which would initiate cascading effects throughout the nuclear sector. The nuclear sector is intertwined with the energy sector because of its production of electricity. In a strange twist, the nuclear sector depends on the energy sector for power, and an interruption to a reliable supply of power would directly affect nuclear facilities in a region affected by a downed electrical grid.

Due to the inherently hazardous characteristics, the nuclear sector depends across the FSLTT echelons of government for emergency response services. The health care and public health sector depends on the nuclear sector for "pharmaceuticals and diagnostic and therapeutic substances" (CISA 2015k). The nuclear sector depends on the health and public health sector for the well-being of its workers. The nuclear sector depends on the information technology sector to "control critical processes, manage day-to-day operations, to store sensitive information, and to facilitate information sharing, including dissemination of security and threat data" (CISA 2015k). The nuclear sector is dependent on the "telecommunications sector for much of its communications capabilities" (CISA 2015k). The nuclear and chemical sectors are mutually supportive as the "principal hazard to public health and safety during an accident at a nuclear facility would be from the release of onsite chemicals or radioactive material" (CISA 2015k).

Chapter 17

Transportation Systems Sector

Transportation systems consist of fixed and mobile assets that touch everyone's lives and provide essential services for the national security and economy. Nearly "9.3 percent of the U.S. GDP was supported by the transportation sector; nearly 6 percent of the 2013 population worked in the sector; in 2014, approximately 19.6 billion tons of goods were transported in the U.S." (CISA 2015l). The mission of the transportation sector is to

> "continuously improve the security and resilience posture of the nation's transportation systems to ensure the safety and security of travelers and goods; it's vision is a secure and resilient transportation system, enabling legitimate travelers and goods to move without significant disruption of commerce, undue fear of harm, or loss of civil liberties." (CISA 2015l)

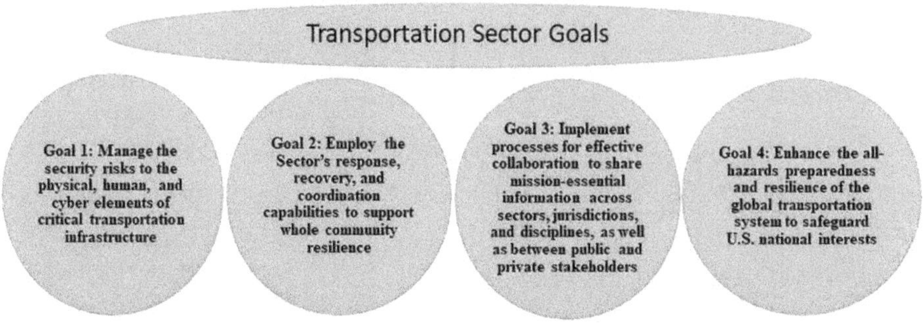

Transportation Sector Goals

Goal 1: Manage the security risks to the physical, human, and cyber elements of critical transportation infrastructure

Goal 2: Employ the Sector's response, recovery, and coordination capabilities to support whole community resilience

Goal 3: Implement processes for effective collaboration to share mission-essential information across sectors, jurisdictions, and disciplines, as well as between public and private stakeholders

Goal 4: Enhance the all-hazards preparedness and resilience of the global transportation system to safeguard U.S. national interests

Transportation systems allow for the lifeline services for small communities up through major metropolitan areas. The sector consists of public and private critical infrastructure owners and operators who share responsibility for managing the security and resilience of the vast network of roadway, rails, and waterways. At the national level, "the transportation sector has four subsectors: aviation; maritime; surface; and postal and shipping" (CISA 2015l). Each contributes to

> "national security, economic stability, public health, and safety. At the regional level, state, and local levels, the sector's effectiveness is determined by businesses, lifestyles, and the emergency needs of the community" (CISA 2015l).

Aviation

The aviation subsector "consists of airports, heliports, seaplane bases, support services, air traffic control, and navigation facilities; there are approximately "19,700 airports in the U.S. with 500 offering commercial service; every month pre-pandemic, approximately 780,000 flights occurred" (CISA 2015l).

Maritime

The maritime subsector is geographically dispersed and consists of "waterways, ports, and intermodal landside connections; there are nearly "95,000 miles of coastline, 361 ports, more than 25,000 miles of navigable waterways, and more than 29,000 miles of marine highways (CISA 2015l).

Freight rail

In 2013, there were approximately "1.33 million freight cars in service and 140,000 miles of active rail track which transported more than 70 percent of all U.S. coal shipments and generated $73 billion in operating revenue for the seven Class 1 railroads" (CISA 2015l).

Highway and motor carrier

This subsector is comprised of "bridges, major tunnels, trucking carrying hazardous materials, other commercial freight services, motor coaches, school buses, and key intermodal facilities; includes nearly four million miles of roadways, more than 600,000 bridges, and 400 tunnels" (CISA 2015l).

Pipeline

There are "more than 2.5 million miles of pipelines spanning the U.S. that transports nearly all of the natural gas and 65 percent of hazardous liquids, including crude and refined petroleum; this subsector includes all above the ground assets include compressor stations and pumping stations" (CISA 2015l).

Postal and shipping

This subsector consists of "large integrated carriers, regional and local courier service providers, mail services and mail management firms, and chartered and delivery services; every day, approximately 720 million letters and packages are transported" (CISA 2015l).

Mass transit

This subsector "includes transit buses, trolleys, monorails, heavy rail, light rail, passenger rail, commuter rail, and vanpools and other forms of rideshare; in 2012, 10.3 billion people moved via mass transit systems (CISA 2015l).

Partners

The transportation sector touches everyone for personal and professional movement and directly enables the daily operations and overall success of ten other sectors of critical infrastructure. The transportation sector aids the "chemical, communications, critical manufacturing, dams, defense industrial base, emergency services, energy, food and agriculture, information technology, and water sectors" (CISA 2015l). As such, the sector is crucial in supply chain operations and for the "transportation of raw materials to factories, the delivery of refined or manufactured products to buyers, and the shipment of agriculture and food products" (CISA 2015l).

Threats to the transportation sector

The risks to the transportation sector include natural disasters, man-made threats, and aging infrastructure. Natural disaster risks include "earthquakes, wildfires, flooding, and extreme weather events, such as blizzards, hurricanes, tornados, and droughts, all of which have the potential for widespread disruption of transportation services" (CISA 2015l).

Man-made threats include "terrorism, vandalism, theft, technological failures, accidents, and cyberattacks" (CISA 2015l). Cyber risks are "an increasing concern as transportation operations become more reliant on information technologies, foreign markets, international mobility, and global supply chains" (CISA 2015l). Terrorist attacks, whether physical or cyber, can disrupt vital transportation services, impact the public's belief in its government's effectiveness, and inflict long-term economic consequences.

The aging and inevitable deterioration of many structures threaten the resilience of these systems and can expand outward and inflict greater damage to regions when combined with man-made or natural disasters. The loss of a key node or asset "poses an immediate threat to users and can have cascading impacts to passenger and freight movement, as well as potentially large-scale impacts such as supply chain disruptions" (CISA 2015l).

Chapter 18
Water and Wastewater Systems Sector

"Drinking water and wastewater treatment are essential to life and the nation's economy; safe drinking water and wastewater treatment is essential for protecting public health, the environment, and preventing disease" (CISA 2015m). There are approximately "153,000 public water systems in the U.S. and are categorized according to the number of people they serve, their source of water, and whether the same customers are served year-round or on an occasional basis" (CISA 2015m). Public water systems are further subdivided as

> "community water systems, which serves people year-round in their residences; non-transient non-community water system, which is not a community water system but still regularly serves at least 25 of the same people more than six months of the year such as schools, factories, office buildings, and hospitals that have their own water systems; and transient non-community water system, which is serves transient consumers who are individuals who have the opportunity to consume water from a water system but who do not fit the definition of a residential or regular consumer, such as gas stations or campgrounds

> where people do not remain for long periods of time." (CISA 2015m)

In the US, there are approximately "51,000 community water systems, 18,000 non-transient non-community water systems, and approximately 84,000 transient non-community water systems" (CISA 2015m).

Water

In the US, water may be sourced from the ground, surface, or a combination of the two. The majority of smaller communities of less than 10,000 people source their water from the ground, whereas larger communities source their water from surface sources. Nearly all water goes to a treatment plant prior to the public where it is conveyed via pipes, open canals via pumps, or it could be gravity-fed. Often, water awaiting treatment is held in reservoirs or lakes which may be in remote or urban areas. Once it arrives at the treatment plant, water may be filtered using a "variety of physical and chemical treatments, depending on the contaminants in the untreated water" (CISA 2015m). Once the water has been treated, it is stored before being distributed to customers either in large covered or uncovered reservoirs. An elaborated network of pipes, tanks, and pumping stations covey water to customers at a carefully controlled and highly precise volume, flow, and pressure. Monitoring for conventional and unregulated contaminants is an ongoing endeavor, which includes continuous readings for the volume, flow, and pressure of the water flowing through the pipes.

Wastewater

A disruption of wastewater treatment can immediately result in a serious public health and environmental impact. It will also cascade quickly and impact the economic well-being of the region, and if wastewater operations cease for any reason, the lack of redundancy in the sector will result in a denial of service. The majority of wastewater in the US is treated by publicly owned treatment facilities, but there is a small number of privately owned treatment facilities in industrial plants. In the US, there are more than "16,500 publicly owned treatment facilities that service the wastewater for more than 227 million people" (CISA 2015m).

Threats to the water and wastewater sector

The risks to this sector are the usual ones highlighted throughout the previous fifteen chapters with the other fifteen sectors of critical infrastructure that are both natural and man-made: natural disasters such as floods, earthquakes, hurricanes, ice storms, pandemics; an aging infrastructure; terrorism; and finally, cyber.

Partners

The water and wastewater sector supplies services and products to the other fifteen sectors. In fact, without water and wastewater treatment, all of the sectors and the economy of the US would come to an abrupt halt. The water and wastewater sector are highly dependent upon the energy, information technology, chemical, critical manufacturing, and health and public health sectors in order for it to operate.

Chapter 19

What Is the Cyber Threat Today

Knowledge is power, but only if you know how to use it, and this statement best describes cyber capabilities and those who develop and use them in the twenty-first century. Cyber-related attacks have increased globally against nations, companies, and individuals.

> [On] November 16, 2021, the House of Representatives Subcommittee on Oversight, Investigations and Management subordinate to the Committee on Homeland Security House of Representatives, convened and received testimony from an extensive list of distinguished people in government and the commercial sector highly knowledgeable of the cyber threat posed to this country. The list included the following:

1. Honorable Michael T. McCaul, Chairman of the subcommittee
2. Richard Clarke, Former Special Advisor on Cybersecurity to President Bush
3. Shawn Henry, former executive assistant Director, Criminal, Cyber, Response, and Services Branch in the FBI

4. James A. Lewis, Director and Senior Fellow, Technology and Public Policy Program, Center for Strategic and International Studies
5. Gregory C. Wilshusen, Director, Information Security Issues, Government Accountability Office (GAO)
6. James R. Clapper, Director of National Intelligence
7. Stuart McClure, Chief Technology Officer, McAfee
8. Stephen E. Flynn, Founding Co-Director, George J. Kostas Research Institute for Homeland Security, Northwestern University
9. John Watters, Chairman and CEO, iSight Partners, Inc. (US Congress, 2012)

According to their testimonies, the threat is real, and attacks are occurring every day against individuals, companies, and multiple echelons of the government, in particular the federal. The cyber culprits are both individuals, groups, foreign government officials protecting the previous entities, and the foreign government themselves, indicating a non-declared war on this nation's national security secrets and commercial interests.

According to the DHS Science and Technology office:

> "cyber adversaries have presented a full spectrum of threats not only to the U.S. government but also to private organizations and critical infrastructure sectors. As the internet grew, criminal, malicious internet activity spread through profit gain, hacktivism, and espionage. Perhaps even more visible than adware and spyware were phishing emails, zero-day attacks, rootkits, and rogue anti-spyware incidents." (DHS, n.d.)

On October 6, 2021, the *Washington Post* held a live virtual cyber conference on cybersecurity. According to Jen Easterly, the CISA director, Russia has continued to engage in state-sponsored or supported cyberattacks against the US and commercial entities despite warnings by US officials, including Biden. Dmitri Alperovitch, chairman of the Silverado Policy Accelerator and one of the participants, stated that "the civility of the old days is dead. The idea that critical infrastructure would be viewed as taboo, hands-off due to their criticality in the daily lives and economic well-being of a country, is over" (Washington Post Live 2021). On August 4, 2021, Eskenazi Health, a hospital located in Indianapolis, Indiana, was forced to power down its network when it found itself the victim of a ransomware attack (Drees 2021). The hospital was forced to divert inbound patients to other hospitals and reverted to "manually writing everything, from prescriptions to lab work, to instructions for home care after surgery" (Drees 2021).

US Army General Paul Nakasone, the director of the National Security Agency (NSA) and US Cyber Command commander, suggested that cybercriminals must be pursued around the globe by law enforcement. It is conceivable that cybercriminal entities may turn on their protectors for many reasons, both personal and professional. Until that occurs, cybercriminals will continue to operate under the protection of government officials that use them for personal gain and to further international intrigue. Currently, Interpol and FBI agents track them and issue warrants for their arrest in absentia, yet there are no incentives nor negative consequences for cybercriminals to cease their illegal activities as long as they continue to receive protection from their respective governments. Unnamed US government officials state that "in Russia, ransomware groups are operating in the permissive environment that they've created there" (Schaffer 2021). In September 2021, FBI deputy director Paul Abbate stated that "the situation is getting worse, much worse" (Schaffer 2021). According to Director Easterly, "Russia has not significantly changed its behavior in cyberspace since President Biden's warning to Russian President Vladimir Putin" (Washington Post Live 2021). The idea "that there's some sort of red line on critical infrastructure does not seem to be holding" (Schaffer 2021), said Dmitri Alperovitch, the chairman of the Silverado Policy Accelerator. "Right now, they have no limits, and part of the issue is that ransomware is not rising to the top of the U.S.—Russia agenda during bilateral discussions" (Schaffer 2021). The Russians do not view ransomware as a critical issue, and the US has not convinced them that it is.

Stuxnet

Surprisingly, few people outside the government and companies associated with cybersecurity know of Stuxnet, yet it is widely viewed as the first modern-day cyberattack against a nation-state, possibly by another nation-state(s) utilizing groundbreaking malware that surprised and stunned the global community. If what many computer experts have pieced together in the aftermath of the attack is true, one could think of it as a non-declaration of cyberwar, which occurred in 2010. Computer experts believed at the time that Stuxnet was the result of the largest and costliest malware development in history (Gross 2011) The malware was so sophisticated that it crossed multiple industrial realms requiring a team of highly knowledgeable, capable, infrastructure-focused programmers in a highly focused and costly effort estimated to have been $100 million (Ryland 2020; Millan 2010; Keizer 2010; Gilbert 2014). Stuxnet had been designed to sabotage the uranium enrichment facility at the Natanz nuclear enrichment laboratories in Iran, which was actively pursuing a nuclear enrichment program to build a nuclear bomb. Bits of the Stuxnet code was later discovered, captured, and analyzed by computer experts. It was determined that multiple safeguards were built into the code, which many interpreted as proof that a Western government had created it (Gross 2011).

Stuxnet was ingenious. In the language of the IT world, it was designed to

> "target supervisory control, and data acquisition (SCADA) systems, which is a control system architecture that comprises computers, networked data communications and graphical user interfaces for high-level supervision of machines and processes." (Boyer 2010)

Linked to SCADA are

> "programmable logic controllers (PLC), which are found in industrial-grade computers that have been adapted to allow for the automation of electromechanical processes such as those used to control machinery and industrial processes; PLC can have thousands of inputs and outputs which are often networked to other programmable controllers and SCADA systems." (Tubbs 2018)

In the case of Iran, these automated systems controlled the operation of the gas centrifuges, which are critical components of the nuclear refinement processes. Stuxnet covertly increased the rotor speed of thousands of uranium centrifuge machines, then lowered the speed without indicating changes on monitoring devices manned by Iranian engineers. The excessive speeds and vibrations, followed by lower speeds, caused the aluminum centrifuge tubes to expand and retract, creating warps that destroyed them (Stark 2011).

Stuxnet quite possibly exploited three zero-days in Siemens's industrial software, in addition to the four Microsoft zero-days (Perlroth 2021). It is a fair assumption that unsuspecting plant workers could have had their cell phones infected with Stuxnet and then carried the virus inside the Natanz facility. There are often Russian, the Democratic People's Republic of Korea (DPRK), or even Pakistani experts who travel to support their nation's stated or unstated global goals or simply to gain monetary rewards by providing industrial assistance in support of aiding in nuclear weapons programs. In 2014, this is exactly what Pakistani Dr. Abdul Qadeer Khan did when he provided technical knowledge and supplied aid to Iran and DPRK in their weapons programs (Collins and Franz 2018). In addition, Russia has been willing to aid in building nuclear reactors in Egypt in 1961, 1997, and currently has four planned for providing electricity to Egypt's growing appetite for electrical power (World Nuclear Organization 2021). Stuxnet was one of the first known viruses capable of "jumping" from one electronic device to others, including computer networks.

The Bush administration is believed to have initiated the effort that created Stuxnet to destroy the fledgling nuclear program in Iran by infecting their computer systems (Kelly 2012). Many foreign government and industry experts believed that the US had been behind the cyberattack against the Iran facility. The remarkable coincidence is that John Bumgarner, a former member of the US Cyber Consequences Unit, published an article outlining a notional cyberattack against a nation's nuclear centrifuges that supported a nuclear program (Carroll 2011). Bumgarner speculated that cyberattacks were legitimate against rogue nations whose uranium enrichment programs "violated international treaties" and stated that the centrifuges used to produce fuel for nuclear weapons could be attacked by cyber warfare. He stated that the centrifuges could be manipulated to cause them to "self-destruct through the manipulation of their rotational speeds" (Bumgarner 2011). In a 2012 interview, former Air Force General Michael Hayden, director of the CIA and NSA, stated that his personal belief that Stuxnet was "a good idea" even though it had legitimized the use of cyber warfare (Kroft 2012). Sean McGurk was a former cybersecurity official at DHS, and his view was that the Stuxnet creators had "opened the box and had demonstrated a new capability," and as a result, "it was not something that could be put back" (Kroft 2012). Both the US and Israel have never publicly commented on the Stuxnet attack.

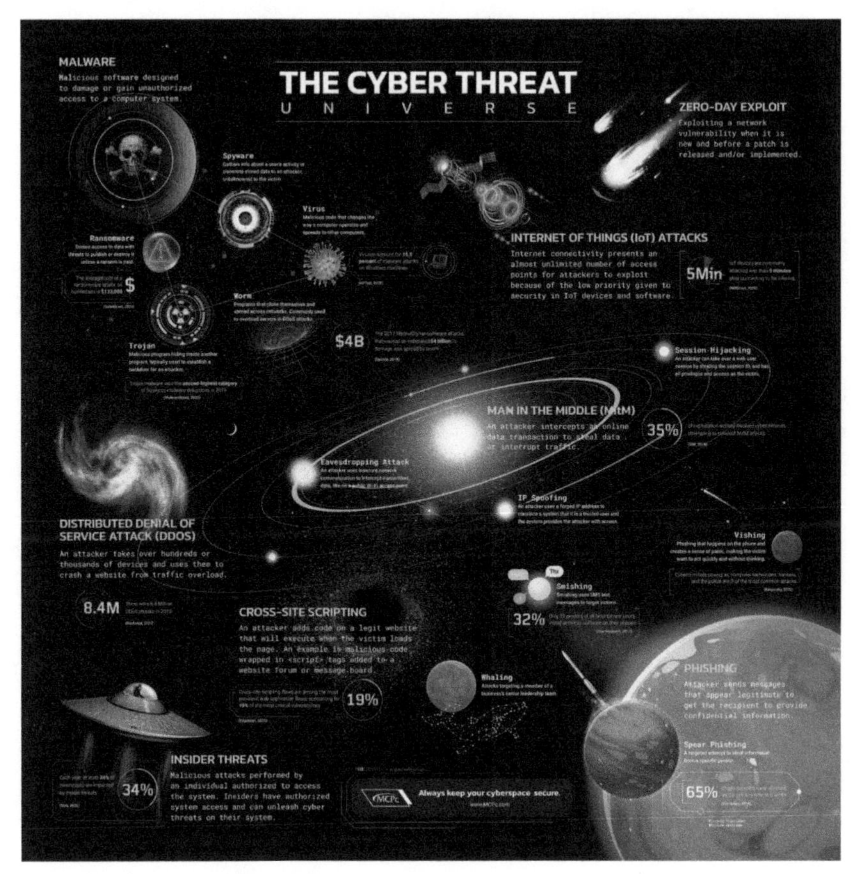

Image courtesy of the Cyber Threat Universe courtesy of MCPc, https:// www.mcpc.com/ Insights /Infographics/The-Cyber-Threat-Universe

Cyberattacks

The number and type of cyberattacks seemingly increase every year, and 2021 was no exception. The term cyberattack refers to "an individual or an organization intentionally attempting to breach the information system of another individual or organization (Lukeheart 2022). The news often features cases in which there is an economic goal involved, but attacks can also be initiated for political, social, and even to copy and destroy an organization's proprietary data. Today there are ten common forms of cyberattacks. They are malware, phishing, man in the middle, denial of service, server query language (SQL) injections, zero-day exploitation, password theft, cross-site scripting, and the Internet of Things attacks.

Malware

The term malware is possibly the most successful cyber tool available to cybercriminals and certainly the most discussed on the evening news because it exploits inherent system vulnerabilities to breach a network. Malware is "any software intentionally designed to cause damage to a computer, server, client, or computer network" (Lukeheart 2022). The attack occurs quite easily: an unsuspecting victim receives an email that seems too good to be true and clicks on a link or email attachment. Instead of receiving $100 million from a Nigerian prince relative, the victim inadvertently installs malicious software inside the computer or network. From there, the malware can deny access to legitimate users, deny access to critical components of the network, retrieve data from hard drives, and spread through the network similar to a virus within the human body; only this malware will cause network disruptions or even halt operations. Malware consists of four major types of cyber programs:

Trojans, worms, ransomware, and spyware

Named after the infamous Trojan horse mentioned in Homer's *Iliad* tale, Trojans are programs that hide within other programs. Trojans do not behave in a manner similar to viruses in that they do not replicate and spread. Trojans find weaknesses and create a back door that hackers can use against a specific target for malicious purposes. Worms are programs that multiply and spread everywhere. Email attachments say those from Nigerian princes offering to share their wealth with long-lost relatives and are surreptitiously downloaded and installed on the victim's computer, then spread to other computers through the victim's contact list and networks through normal email exchanges and email attachments. Recently, worms have targeted email servers and created a denial of service (DoS).

Ransomware has become famous as of late with several high-profile, well-publicized attacks. Ransomware infects networks and denies the victim access to their data. Commercial secrets are now the most fleeting in today's virtual-based, cyber world. Cybercriminals will use ransomware to threaten companies with the public release of embarrassing information, emails, files, etc. or delete critical files unless the victims pay a ransom. Advances in ransomware have evolved into crypto viral extortion, which encrypts a victim's data on their network, thereby making it impossible for the victim to access their data without a decryption key. Lastly, spyware programs are installed to collect information about users and report this information. The attacker can use this information for commercial purposes, where the victim shops, browse, interests, vices, or cybercriminals can use it to blackmail victims while spreading spyware throughout the contacts list to other computers, networks, and victims.

Phishing

Phishing attacks are launched every day across the globe in mass fraudulent emails disguised as reliable sources. The emails appear legitimate, but once the email is opened or the attachment is clicked on, a malicious file or script is installed that allows the cybercriminals access to the victim's device. Once installed, cybercriminals can gather information and spread it to other networks. Social networks have become the favored site for launching phishing attacks as the youth are often naive and freely share personal information. By focusing on social networks, cybercriminals can easily collect information about their interests, activities, location, and work, which grants attackers insight that allows them to focus their criminal enterprise against a specific group, company, or government agency.

The three main phishing attacks are spear phishing, whaling, and pharming. Spear phishing refers to targeted attacks against specific companies or individuals. Whaling is an attack focused on senior executives or stakeholders within an organization. Pharming involves a highly technical attack in which a domain name system (DNS) caches are manipulated, and the attacker can capture a targeted user's credentials, often through a fake login page. Technological advancements now allow phishing to be launched via a phone call, called voice phishing, and via text messaging, called SMS phishing.

Man-in-the-middle (MitM) attacks

MitM refers to a cybercriminal interception of a message between two parties and inserts him or herself into the message. Once inserted, cybercriminals can copy and steal data and manipulate the traffic. This is the preferred method of attack for hackers looking to exploit security vulnerabilities in a network, including those readily available at many unsecured public Wi-Fi networks. The average user would never know that they have been hacked as all the information appears to be moving normally in a legitimate manner between them and their intended destinations. Sometimes, phishing or malware attacks can be used initially to leverage MitM attacks. Cybercriminals have become more efficient by inserting ransomware worms directly into networks.

Denial of service (DOS)

DOS attacks "work by flooding systems, servers, and networks with routine traffic to overload resources and bandwidth" and launch from a specific location (Lukeheart 2022). If successful, this results in overwhelming the system until it cannot process requests. There are also "distributed denial-of-service (DDoS) attacks launched from infected host machines and multiple locations using bots and zombie computers" (Lukeheart 2022). When a DOS takes a network offline, a DDoS attack creates an opportunity for different cyber tools to enter the network and initiate a different attack, then or later. In Britain, the National Crime Agency "discovered that the number of DDoS attacks against school networks had doubled from 2019 and 2020" (Cluley 2022) and was curious as why this had occurred. The investigation revealed that school-age kids ranging from nine to fifteen had discovered how to launch DoS and DDoS from playing computer games against their friends. As their cyber skills improved, so did their confidence, and they then decided to launch the same types of attacks against their school networks for a variety of personal reasons and gain.

SQL injections

A SQL injection intends to insert malicious code into a server to "retrieve protected information; this attack usually involves submitting malicious code into an unprotected website comment or search box" (Lukeheart 2022).

Zero-day exploit

A zero-day exploit exploits a network vulnerability before a patch is released or implemented. When a tech company finds such a flaw in its software or hardware, it has zero days to fix it or suffer the consequences. Zero-day attackers jump on discovered or, worse, announced vulnerabilities when no preventative measures or fixes are in place.

Password attack

Passwords remain "the most widespread method of authenticating one's identity to access secure information systems, making them a highly attractive target for cybercriminals" (Lukeheart 2022). By stealing passwords, cybercriminals sell or distribute them on the dark web or use them to access personal accounts, confidential data, and network systems. Cybercriminals can use several methods to identify passwords, including

> "social engineering, gaining access to password databases, testing network connections to obtain unencrypted passwords, or creating a program designed to try all the possible variants and combinations of information available to a targeted individual to guess the password." (Lukeheart 2022)

This is referred to as a brute-force attack. Another common method is the dictionary attack, when the attacker uses a list of common passwords to access a user's computer and network.

Cross-site scripting

This type of attack "sends malicious scripts into the content available from reliable websites; the malicious code is enjoined to the dynamic content and sent when information is requested from a browser" (Lukeheart 2022).

Rootkits

Rootkits are "installed inside legitimate software to gain remote control and administration rights over a system where it allows the cybercriminal to steal passwords, keys, credentials and retrieve data" (Lukeheart 2022). Rootkits

> "hide in legitimate software and once you allow the program to change one's operating system, the rootkit is installed in the host computer or server system; the rootkits remain dormant until activated and are commonly spread through email attachments and downloads from insecure websites." (Lukeheart 2022)

Internet of things (IoT)

It is difficult to imagine a world before the Internet. This is due to the DOD funding of the Advanced Research Projects Agency Network (ARPANET) in 1960. This crude prototype would evolve into the modern-day Internet. On October 29, 1969, ARPANET sent its first message, a node-to-node transmission from one small house-sized computer at UCLA to another at Stanford (Andrews 2019). The message, "LOGIN," was simple; unfortunately, the Stanford computer only received the letters LO before the primitive network crashed (Andrews 2019).

Technology continued to evolve, and in the 1970s, after scientists Robert Kahn and Vinton Cerf developed Transmission Control Protocol and Internet Protocol (TCP/IP), a communications model that set standards for how data could be transmitted between multiple networks; ARPANET incorporated TCP/IP on January 1, 1983, and in 1990, Tim Berners-Lee invented the World Wide Web (www), which became the modern Internet (Andrews 2019). The www is often referred to as the Internet, but the www is the most common means of accessing data online. Wi-Fi allows users to easily connect to the Internet regardless of where one might be. Wi-Fi is not an abbreviation but a marketing term used to "connect computers, smartphones, and other devices to the Internet; Wi-Fi is a radio signal that is sent from a wireless router to nearby devices" (Andrews, 2019); they, in turn, translate the signal into data. The Wi-Fi-enabled device will transmit a radio signal back to the Internet-connected router (Andrews 2019).

It should come as no surprise to anyone who has purchased any modern device such as kitchen appliances, rechargeable toothbrushes, and nearly any other device that all of our modern devices have Internet connectivity via Wi-Fi or Bluetooth built into them. Thus, these manufactured products are now connected to, controlled by, or share data with companies and other devices via the Internet. Companies purport that they have designed their products to maximize convenience for the customer, although the average customer did not ask for nor want Internet connectivity built into their toothbrush. Interestingly, the convenience has a downside; it has also opened a vulnerability for customers. The *interconnectedness* of IoT in all modern devices offers unlimited exploitation points for the cybercriminal to hack into one's home or business network and spread onto other devices on the network and to other networks. It should come as no surprise that due to the rapid expansion of IoT devices, IoT attacks are increasing due to inherently poor security protocols. One less publicized IoT attack involved accessing an internet-connected thermometer inside one fish tank to access the casino's computer network (Wei 2018). According to Nicole Eagan, the CEO of cybersecurity company Darktrace:

> "the hackers exploited a vulnerability in the thermostat to get a foothold in the network. Once there, they managed to access the high-roller database of gamblers and then pulled it back across the network, out the thermostat, and up to the cloud. There are a lot of IoT devices, everything from thermostats, refrigeration systems, HVAC [air conditioning] systems to people who bring in their Alexa devices into the offices. "There's just a lot of IoT. It expands the attack surface, and most of this isn't covered by traditional defenses." (Wei 2018)

Famous 2021 attacks

Cyberattacks have proliferated and have increased in their complexity. The evolution of technology has extended to cyber platforms, customizable for every imaginable attack purpose. In 2021, cyberattacks became more frequent and disruptive than previously. They also became more profitable for the cybercriminal organizations responsible for them. Since 2020, ransomware has increased 62 percent globally. This success will only encourage cybercriminals. Here are the top ten known cyberattacks of 2021.

The Colonial Pipeline. In May 2021, a cyberattack was launched against the Colonial Pipeline, the largest fuel pipeline in the US. The attack disrupted fuel deliveries in twelve states, resulted in chaos across the East Coast, and lasted several days. The attack was the result of a single compromised password, according to a cybersecurity consultant who responded to the attack (Turton and Mehrotra 2021). Cybercriminals gained access into the Colonial Pipeline Co. network remotely through a virtual private network account. Colonial "paid the hackers, an affiliate of a Russia-linked cybercrime group known as Dark Side, a $4.4 million ransom shortly after the hack; the hackers also stole nearly 100 gigabytes of data from Colonial Pipeline and threatened to leak it if the ransom was not paid" (Admin 2021).

JBS. Brazil's JBS, "the world's biggest meat processor, suffered a cyberattack that resulted in the temporary closure of operations in the U.S., Australia, and Canada" (Admin 2021). The attack "threatened supply chains and caused further food price inflation in the U.S.; to prevent further disruptions, JBS paid the $11M ransom" (Admin 2021). The FBI described the criminal group responsible as one of the most specialized and sophisticated in the world.

Florida's water supply. In a bizarre attack, "a cybercriminal managed to infiltrate the city of Oldsmar's computer system and, for a brief time, increased the sodium hydroxide level in the water supply to dangerous levels" (Admin 2021).

Australia's Channel Nine. The Australian Channel Nine station was targeted in an attack that prevented the channel from airing its Sunday News bulletin plus other shows. The incident occurred "simultaneously as a suspected attack on Australia's parliament in Canberra, triggering concerns about the country's vulnerability to cyberattacks in general" (Admin 2021).

CNA insurance. CNA, one of the larger insurance firms in the US, was the victim of a cyberattack that caused it to cease trading. The breach caused network disruption and impacted email. Third-party forensic experts "determined that an updated version of the Phoenix Crypto Locker Malware, a form of ransomware, was used" (Admin 2021).

Microsoft Exchange Server attack. By exploiting Microsoft's Exchange Server vulnerabilities, this cyberattacks affected millions of Microsoft users globally. In the US, an estimated sixty thousand private companies and federal and state governments suffered disruptions.

Bombardier data breach. In February of 2021, the aerospace company "Bombardier suffered the compromise of confidential data involving 130 employees plus information, customers, and suppliers; the vulnerabilities in their third-party File Transfer application were found to be to blame" (Admin 2021).

Acer ransomware attack. Acer, a computer hardware manufacturer, suffered "a security breach resulting in a ransom of $50 million; the attack is believed to have been carried out by cybercriminal group, REvil, who later leaked some of the stolen data online" (Admin 2021).

University of the Highlands and Islands. Scotland's University of the Highlands and Islands was "forced to close all of its colleges and research labs to students due to a cybersecurity incident. The attack was notable as it utilized a penetration testing toolkit known as Cobalt Strike, normally used for legitimate purposes" (Admin 2021).

The Accellion supply chain attack.

> "Confidential data was stolen from several large organizations like Singtel, The University of Colorado, and The Australian Securities & Investments Commission when security software company Actelion's File Transfer System was breached and subsequently leaked online." (Admin 2021)

The Chinese have hacked US intellectual properties for decades. Iran and the DPRK both have demonstrated their capabilities: Iran spoofed a high-tech US UAS, causing it to land undamaged inside of Iran; later it caused a crash of US banking websites networks in Las Vegas; the DPRK attacked Sony servers in retaliation for a movie that made fun of Dear Leader Kim and then hacked into a Bangladesh bank and stole $81 million (Perlroth 2021). So far in 2021, Ransomware has surpassed the other five categories of cyber threats in an increased number of attacks, success, and devastation. In July 2021, a ransomware attack was linked by the FBI to cybercriminals in Russia that created turmoil from Swedish supermarkets to kindergartens in New Zealand (Kim 2021). Irina Borogan, a senior fellow with the Washington-based Center for European Policy Analysis, says computer experts are sure that the criminals are Russians, though they lack direct evidence. Borogan stated that the Russian authorities are protecting the criminals possibly due to financial and international political gain. These same Russian cybercriminal organizations were blamed for the massive ransomware attacks earlier in 2021 on the Colonial Pipeline and meat processor JBS (Kim 2021).

In order of their open-source perceived capabilities, the Russians, Chinese, Iranians, and even the DPRK have extensive offensive cyber capabilities. The joke inside the DoD for many years was that the DPRK has only two computers: one for dear leader Kim and one for the rest of the government. When Kim Jong-il met with Secretary of State Albright in 2000, she was surprised when at the end of what had been described as a very productive two-day series of meetings, he turned to her and famously asked Albright for her e-mail address (*The Economist* 2007). This simple request surprised everyone in the state department. The server that Kim used was based in China as the infrastructure within the DPRK, at least at that time, could not support Internet requirements. That has changed, and in many circles, the DPRK hackers are perceived as being in the top five of the nation's supporting or sponsoring active global cybercriminal activities. They have mastered the lessons from their former cyber masters, the Chinese. It is purported that "hackers with suspected links to the Pyongyang dictatorship have been going after Chinese security researchers in an apparent attempt to steal their hacking techniques and use them as their own" (Vavra 2021). The Chinese government supported hacking efforts, and the DPRK hackers were intent on stealing government and commercial secrets and then copying the hacking tools for their own uses. Keith Alexander, the former director of the NSA, called "Chinese cyber espionage the greatest transfer of wealth in history" (Perlroth 2021). In an effort to solidify the legacy of President Xi, the Chinese are openly attempting to surpass the US as the rising superpower to contend with in the burgeoning twenty-first century. Cyber intrusions are easy to launch from distant locations. Unmanned systems can be used as vehicles to launch or augment cyberattacks and can even be used in direct support or in supporting attacks.

Stuxnet became the new form of warfare in the twenty-first century: sneaky, invisible, anonymous, and potentially devastating to a targeted nation's industrial complex, its defenses, and its very dependence on all things digital and computerized. It is scary to contemplate where capabilities have evolved since 2010. The Russians, in their multiple recent invasions of Georgia and Ukraine, used cyberwarfare to disrupt, deny, and dissuade the Georgian and Ukrainian militaries long before the first Russian troops and tanks crossed the border.

> "Cyber operations attributed to Moscow are not conducted in a strategic vacuum. They are enabled and shaped by broader geopolitical considerations and the institutional culture of Russia's military, intelligence, and political leadership, as well as by Moscow's evolving approach to asymmetric interstate competition that falls short of all-out conflict. Russia's conceptualization of warfare has shifted from a consensus that the baseline of warfare is armed violence to an agreement that the baseline for warfare has broadened to include a tailored amalgamation of armed violence and non-military measures." (Chekinov and Bogdanov 2015a; Chekinov and Bogdanov 2015b; Jonsson 2019; Gerasimov 2013)

Cyberspace recognition

Cyberspace is in desperate need of official recognition and designation as the seventeenth sector of critical infrastructure to our nation. Many senior government officials with whom I have talked have said that cyber is already an important part of their business, their critical infrastructure sector. That may be true, but there is an increased emphasis, acknowledgment, and resourcing to be gained by designating cyberspace as a stand-alone sector of critical infrastructure. It has been demonstrated all across the globe that cybercriminals, or nation-states, have used cyber as a means for inflicting pain and gaining monetary rewards from their sworn enemies. We find ourselves in a state of undeclared war.

The Russian military forces demonstrated in Georgia in 2008 and Ukraine in 2014 that the initial wave of offensive operations occurred prior to the introduction of military forces and involved a cyberattack against both nations' infrastructure, to include all basic aspects of daily lives such as telecommunication systems, traffic systems, and banking systems. The mobilization of defense forces and the call-up of reserve forces were negatively affected by this asymmetric attack. Citizens in Georgia and Ukraine were forced to seek shelter and remain in their homes. Military reservists did not know where to go and how to get there—the trains and street signals were nonoperational. Phones, computers, emails did not work, thus the country's government was cut off from the outside world and from its citizens inside the country. With what the nightly news has reported being a buildup of Russian military forces estimated at 130,000 soldiers or more poised outside the border with Ukraine, it is only a matter of time before Putin gives the order to launch Russian forces into Ukraine, only this time they will not depart until they have overthrown the legitimately elected government and replaced it with a sympathetic puppet regime. While the North Atlantic Treaty Organization (NATO) contemplates its next steps, if any, Russia is undeterred and moving forward with what Putin has no doubt authorized weeks or months ago. If Russia does launch an invasion, we need to pay attention to the opening salvo of the war when Russia launches a herculean cyberattack against all aspects of Ukrainian critical infrastructure, government, commerce, and daily life for the citizens prior to moving its military forces across the border and into Ukrainian territory. It did this in 2008 when it defeated Georgia in twelve days and again in 2014 when it invaded Ukraine and annexed the Crimean Peninsula. But that was not the end of it for Ukraine; sadly, it was only the beginning.

Russia's fingerprints

Putin, the former KGB colonel, president, then prime minister, then president, learned his sneaky tricks working for the notorious organization. Putin was a KGB operative, and the center of his foreign ambitions was to undermine the West across the globe. His hackers would spread lies and disinformation, thus Russia's opponents would be on the defensive in their own countries, battling the political intrigue that his group had created. They would be distracted and ripe for fracture. Putin's real aim was to destroy NATO, the only obstacle to his ambitions to dominate Europe. He learned that you never just beat your opponent; you annihilate him with all the tools available to you in your proverbial tool kit. It has been reported that Putin laid down only two rules for Russia's hackers: first, no hacking inside the motherland; second, when the Kremlin calls in a favor, you do whatever it asks, otherwise the hackers had full autonomy. And oh, how Putin loved them. He described Russian hackers "like artists who wake up in the morning in a good mood and start painting" (Perlroth 2017). In 2017, Putin stated that Russian hackers, "if they have patriotic leanings, they may try to add their contribution to the fight against those who speak badly about Russia" (Perlroth 2021).

While the suspected US Stuxnet attack against the Iranian centrifuges was a surgical strike, the longest continuous and most devastating cyberattack against an entire nation and its citizens must be the Russian's six-year cyber onslaught against Ukraine. In 2016, less than two years after the Russian puppet government was ousted and replaced by a democratic one, Ukrainians were still suffering through the lingering consequences of the Russian led cyberattack against "government agencies, transportation systems, ATMs, gas stations, the postal service, even the radiation monitors at the old Chernobyl nuclear site" (Perlroth 2021). The Russian-built malicious code left Ukraine, and thanks to the interconnectedness of our world via the IoT, it spread and infiltrated computer networks of governments and businesses around the globe.

Ukraine had angered Putin when they kicked out their Russian-installed government in 2014 and held free elections to select their own leaders. Thus, Ukraine became a test bed for Russian hackers. To initiate unrest and to make Ukrainians question the power of their government, he had his hackers initiate outages with the heating companies, the media, the power companies, and the communications networks. A year later, Putin directed his hackers to do it again. Even in 2016, US intelligence assumed that as bad as the Russian hacking was, it was nowhere near what their own capabilities. Sometime within the year, the NSA had cyber tools stolen in what may have been the largest and most successful hack in human history. A group referring to itself as the Shadow Brokers began to sell or give stolen NSA cyber tools to any nation or cybercriminal to use for their own purposes. In 2017, Russian used the NSA cyber tools to once again attack Ukraine in what became the most destructive and costly cyberattack in world history (Perlroth 2021). Ukrainian citizens could not use any electronic device for any simple daily task, and even the radiation monitors at the infamous Chernobyl power plant were taken offline. The attack spread worldwide and even attacked companies inside Russia, including those billionaires that were allied with Putin. It was clear to the NSA that one of their cyber tools had been used in the attack, and in the end, the costs of the single attack were estimated to have been greater than $10 billion. The extent of the damage in Ukraine had been stifled by the fact that the country had not yet joined the IoT.

It is believed that a group of Russian hackers hacked into the American tech company Solar Winds and inserted malware into their codes (Tepperman 2021). No one noticed, and when the next software update was released, the virus went into the networks of more than eighteen thousand clients, including the DoD, Department of State, Department of the Treasury, the NSA, and many others. It was only discovered months later that some of those clients reported vast amounts of their classified data had been stolen (Tepperman 2021). In the past decade, US businesses have incurred billions of dollars in losses due to cyberattacks. It appears that deterring cyberattacks is exceedingly difficult, and the more a country is tied into the IoT, the easier it is to attack that country and its subordinate entities, companies, and individuals who are likewise players in the IoT. The US created a lucrative market for digital weapons and for hackers to find and weaponize vulnerabilities or gaps in existing software.

Cyber and unmanned systems working together

Micro UAS can be employed as easily as someone walking down a hallway and dropping tiny UAS along the way, only this would sit unobserved as it quietly conducted surveillance or, worse yet, performed what is referred to as a vampire tap. In simple terms, "a vampire tap is a cable connector that clamps onto and bites into the cable, hence the vampire's name, and a favorite for MitM hackings" (Rohit 2021). Vampire taps were "used to attach earlier thick Ethernet transceivers to the coaxial cable; without a vampire tap, the cable had to be cut and connectors had to be attached to both ends (Rohit 2021). There are many additional options open to those informed specialists working in the cyber arena to utilize unmanned systems to aid in the surveillance, reconnaissance, and launch of cyberattacks and in the support of said attacks.

DJI and the Chinese threat to US national security

For years, one of the criticisms of DJI was that its UAS covertly send recorded information back to China. Many of the various departments and agencies of the US federal government had used DJI UAS for years. A worst-case scenario would be a federal department using DJI UAS and then taking the UAS back to the operations room in the headquarters for an after-action report. The UAS sits atop a cabinet or in a corner while its battery is recharging. Meanwhile, it is recording and sending the information it has gathered back to China. In past conflicts, countries would send in covert personnel to conduct reconnaissance of key bases, structures, facilities for targeting. Imagine having the enemy use your nation's commercial UAS products to surreptitiously provide them for you while making a profit in UAS sales. As mentioned in chapter 10, "article seven of China's National Intelligence Law passed in 2017, requires organizations to assist in espionage and to keep those activities secret" (Girard 2017).

According to Andrew Shelley, "for the average recreational user who might be taking selfies on the beach, it is probably true that DJI is not interested in their data, but collectively, the Chinese government is interested in our data. We don't understand just how much of a threat that is" (Einhorn and Shields 2021). According to the Department of Interior (DOI), they had an inventory of 811 UAS, most of them made by DJI. DOI used the UAS to cover bridges, monuments, dams, and other infrastructure, some of which are certainly critical that would be a national security threat and certainly useful to Chinese intelligence agencies. Former senior Obama and Trump administration officials have sounded the alarm on the amount of personal data that China is gathering from US citizens, as well as mapping the US courtesy of the recordings made by DJI UAS and shared with Chinese intelligence (Fink 2014; Venable and Ries 2021). Yale Law School professor Oona Hathaway stated that "each new piece of information, by itself, is relatively unimportant but combined, the pieces can give foreign adversaries unprecedented insight into the personal lives of most Americans" (Einhorn and Shields 2021).

According to Matt Pottinger, a former US deputy national security adviser, Chinese President Xi has plotted against the West by aiming to gain data for the purpose of gaining an economic and military advantage. "If Washington and its allies don't organize a strong response, Mr. Xi will succeed in commanding the heights of future global power" (Einhorn and Shields 2021). The competition between the US and China has the potential of shaping the economy of the world for decades. Today virtually all modern-day products are connected to the Internet, and all transmit or receive data. Harvesting all the data that one nation can gather on another will give it a significant advantage. According to another US government official, Paul Triolo:

> "Data security will be a defining issue for the next decade. The democratic and authoritarian digital

> worlds will be built on largely different hardware, with different standards, and limited points of connection, which will drive up costs for businesses operating across these two spheres, reduce innovation, and lead to geopolitical tensions, reduced trade, and a much more complex world for companies to operate within. Other countries will be forced to choose sides in this divide, and this will be painful and costly." (Einhorn and Shields 2021)

In 2019, Trump prohibited the DoD from purchasing DJI UAS, and a year later, "the DOC placed DJI on its Entity List, which bars U.S. suppliers from selling to it without an exemption" (Einhorn and Shields 2021). In 2021, Biden blocked US investment in the company (Venable and Ries 2021). There is a bill before Congress that would ban the federal government from purchasing DJI UAS.

Global threats—2022

According to the London-based think tank the International Institute for Strategic Studies:

> "Russia and China have each dedicated significantly more military cyber forces to conducting cyber effects than the U.S.: 33% of Russia's military cyber forces are focused on effects, compared to 18.2% of Chinese military forces and 2.8% of U.S. forces." (Pomerleau 2022)

They further defined effects as

> "actions to deny, degrade, disrupt or destroy as well as those conducted by proxies in conjunction with a government actor. It can also include a range of other capabilities such as the ability

> to research vulnerabilities, write or use malware, and maintain command and control through exploits." (Pomerleau 2022)

Russia allocates "a significant amount of personnel to incident response, with 80% of its forces dedicated to the mission; for the U.S. the percentage is 29%; for the Chinese it is 9.1%" (Pomerleau 2022). Russia, China, and the US all dedicate approximately 50 percent to 54 percent to conducting cyber intelligence, surveillance, and reconnaissance (Pomerleau 2022).

The threat of a Russia invasion of Ukraine and the threatened economic sanctions that the US has promised to immediately place upon Russian president Putin and his inner circle of oligarchs have resulted in a threat of cyberattacks against the US by the Russians. As a result, the US government's Cybersecurity and Infrastructure Security Agency (CISA) has issued a nationwide warning to government departments, businesses, and other critical organizations to be on their guard against cyberattacks. Cyberattacks became an integral component of Russian invasion plans dating back to its brief foray into Georgia in 2008. CISA further clarified that the US homeland "is not now facing any specific, credible threats," but due to US promises of immediate economic sanctions, "Russia could escalate its destabilizing actions in ways that may impact others outside of Ukraine" (Losey 2022). According to CISA, it has become the aim of the Russian government military operations as

> "the Russian government understands that disabling or destroying critical infrastructure—including power and communications—can augment pressure on a country's government, military and population and accelerate their acceding to Russian objectives." (Lacey 2022)

The FBI revealed that "as of November 2021, Black Byte ransomware had compromised multiple U.S. and foreign businesses, including entities in at least three US critical infrastructure sectors, government facilities, financial, and food & agriculture" (Gatlin 2022). "Black Byte is a Ransomware as a Service (RaaS) group that encrypts files on compromised Windows host systems, including physical and virtual servers" (Gatlin 2022). The National Football League's "San Francisco 49ers was also struck by a Black Byte ransomware attack that stole data from their servers and subsequently leaked 300MB of files" (Gatlin, 2022).

Cyberattacks generally hold a victim's system hostage by encrypting its data until their demands are met, but according to the FBI, "most ransomware incidents against critical infrastructure affect business information and technology systems, but several ransomware groups have developed code designed to stop critical infrastructure or industrial processes" (Baksh 2022).

According to a joint FBI and US Secret Service advisory:

1. "In May, after Colonial Pipeline paid ransomware attackers $5 million to release their system, the company said it had proactively disconnected the operational technology—think valves, and pressure gauges—that control its physical processes. There was no evidence the hackers got beyond their information technology realm.
2. As operational technology and IT have become increasingly intertwined to create greater efficiency of industrial control systems, the sector has become a frequent target of ransomware perpetrators and other malicious actors. One high-profile example of the kind of damage cyber adversaries can do by manipulating OT came last year when an attack attempted to change the chemicals in a Florida water treatment plant to levels that would be unsafe for the community it serves.
3. The ransomware business model that has allowed for large scale commoditization of exploits has only continued to advance over last year.

4. The market for ransomware became increasingly 'professional in 2021, and the criminal business model of ransomware is now well established. In addition to their increased use of ransomware-as-a-service, ransomware threat actors employed independent services to negotiate payments, assist victims with making payments and arbitrate payment disputes between themselves and other cyber criminals.
5. That could be especially bad news for owners and operators of slightly smaller critical infrastructure sectors that rely on industrial control systems. The advisory notes—at least in the U.S.—ransomware perpetrators appear to be staying away from "big game" targets like Colonial Pipeline after U.S. authorities subsequently disrupted their operations.
6. The FBI observed some ransomware threat actors redirecting ransomware efforts away from big-game' and toward mid-sized victims to reduce scrutiny." (Kadsh 2022)

Chapter 20
Unmanned Systems

UAS are not a new entity, though, from the slow reaction and even slower speed at which the FSLTT levels of governments are moving to counter these threats. It has been so slow that one would think that UAS had just been unveiled to the public. The federal government lead for UAS, the FAA, has the unenviable position of regulating our airspace, and the FAA has determined that UAS are legally entitled to the same protection afforded to military, civilian, and commercial aircraft (FAA, n.d.). The FAA is busy developing solutions geared to the mass of law-abiding people, not bad actors. Software and hardware designed for geofencing—warnings to stay out of the area you are flying toward—and digital license plates—this is who owns or is operating this UAS—and remote identification—this UAS is registered to this person—only work if the systems are enabled and the UAS are being operated by law abiding people (Edwards 2021). The bad actors will simply disable the systems or turn off the safety systems, or if they are building their own UAS based on the availability or parts worldwide, they will simply not install the systems required by the FAA. The owner-operators of much of the US critical infrastructure facilities are precluded from procuring sensors to detect UAS and the C-UAS equipment necessary to protect their facilities.

In 2019, the NRC completed a three-year security review, and their determination was that UAS do not pose a threat to nuclear power plants, therefore no authorizations for sensors or C-UAS equipment were required by the private owner-operators of the facilities. In 2014, the French nuclear power plant was "attacked" by two UAS when one stopped at the perimeter and filmed the other flying into one of the reactor buildings; in 2019, the Saudi oil field and refinery were damaged, and the global oil supply was affected by the attack by cruise missiles carrying multiple UAS armed with explosives; in 2020, an electrical power station in Pennsylvania was targeted by a UAS with ropes tangling and copper on the ends which were intended to instigate an electrical short at the facility ad to render it nonoperational. These three examples of critical infrastructure being targeted and attacked are the only ones of which we are aware. There is no telling how many there have been and are being kept from the public. The authority's rationale would be that the release on information of the actual numbers and incidents would only instill fear in the public and in the government's abilities to protect them and the facilities upon which we all are dependent.

As scary as that proposition might be, it will only worsen when bad actors move their destructive efforts into USS and the UUS realm. UAS, and soon the USS and UUS platforms, are capable and have been instruments for cyberwarfare and are threats to industrial, commercial, and personal data. The malicious use of UAS as cyber platforms by bad actors or nations can no longer be ignored as they are upon us. The evolving world of the IoT has migrated to UAS: the easy use, integration to the IoT, and its connectivity have created airborne routers, which, with exceedingly small and lightweight computing devices called Raspberry Pi, can easily use spoofing techniques to exploit unsecured networks and devices. The UAS can make one believe that you are your network, but instead, the UAS has hijacked what you believed was your connection, and you are now communicating through the onboard computer on the UAS. While operating in the air, UAS can hack anyone's smart devices through their built-in Wi-Fi and Bluetooth connections. This now allows UAS to access your personal data and anyone else's.

UAS

In the past decade, the use of UAS in our daily lives has altered society in a way that few envisioned. UAS was born out of military purposes yet has transformed the civilian sector due to their low cost, compared with manned aircraft performing similar surveillance or reconnaissance by federal, state, and local echelons of government. There is a downside with many new inventions, such as automobiles, aircraft, and the Internet, to mention but a few. With the advent of UAS, security, safety, and privacy issues have arisen. These have manifested through aerial bombardment, surveillance, cyberattacks, flying in restricted airspace shared by manned aircraft, and performing illegal surveillance of individuals.

Globally, over ninety countries and non-state actors have developed UAS technologies and use them. A small number of those countries and all the nation-states promote terrorism and utilize UAS to help them achieve their goals. The majority of the UAS systems are low technology systems designed for intelligence, surveillance, and reconnaissance (ISR) purposes, whereas the systems being developed or those already in use by the US, China, Russia, Turkey, and Iran platforms are far more deadly, advanced, and capable of far more than simply ISR (Boutros 2015). Iran has exported some of its UAS to Syria, Sudan, Hezbollah, Hamas, and ISIS to further its foreign policies and spread its ideologies. Israeli intelligence estimates that Hezbollah has an inventory of 200 UAS (Zwijnwenburg 2014).

CISA has organized the UAS threat environment "into three threat groups: the careless and clueless recreational drone user; intentional operators and activists; and terrorists and paramilitary users" (CISA, n.d.b.). CISA analysis of UAS incidents based on reporting has the careless and clueless group as the culprits in most reported incidents. This group prefers commercial off-the-shelf (COTS), multi-rotor UAS, and unwittingly violate the laws and regulations governing the operation of an aircraft. It is important to remember that according to the FAA Modernization and Reform Act of 2012, UAS are classified with the same legal safeguards as that of fixed-wing or rotary aircraft (Federal Aviation Administration 2020a). The intentional and activist users flaunt the law and use modified COTS UAS to drop illegal drugs, money, cell phones, and weapons into prisons and across international borders, such as the southwest of the US, where the Customs and Border Patrol (CBP) is in a constant battle with Mexican drug cartels. Lastly, the terrorists and paramilitary groups are idealists intent on creating havoc. This group has money, purchases larger, faster custom UAS with longer ranges and improved software and increased lift capabilities. For their intended purposes, these UAS are a small investment with the potential for a huge payoff, should their illegal activities come to fruition. Their goals may be to garner media attention with a political message or initiate an autonomous attack while they are far from the scene.

CISA conducted a study and identified five tactics, physical and cyber. The physical tactics include "weaponization, smuggling, disruption or harassment, and surveillance or reconnaissance" (CISA, n.d.b). Weaponization encompasses one or more modifications to a UAS to aid in an attack or inflict increased casualties, damage, and fear. An example of this would be the rapid development of a plastic tube that contained an explosive and release mechanism by ISIS to drop explosives on soldiers in the open or on vehicles in Syria. Smuggling consists of a UAS to airdrop contraband materials. Two modern-day examples of this are the widespread use of UAS to deliver restricted and illegal items, including drugs and weapons, into prisons and the intentional dropping of bundled drugs across the US border by Mexican drug cartels.

Disruption is an intentional act to harass or hinder the security and operators of critical infrastructure facilities, unique events such as sporting events, or individuals and manned aircraft by bad actors utilizing UAS. An example of this occurred in 2018 and involved using a small commercially available UAS that shut down operations at Gatwick Airport near London (British Broadcasting Corporation 2018a). This intentional act resulted in the cancellation of eight hundred flights that affected over one hundred thousand travelers (Mueller and Tsang 2018). The shutdown cost the government, airports, airlines, and passengers an estimated $65.5 million (Calder 2019).

Reconnaissance and surveillance involve using the UAS camera to surveil law enforcement manning and operating procedures or security personnel at critical infrastructure facilities. It can also include economic or industrial espionage. Reconnaissance is often done during the initial planning stages of an attack, while surveillance is done during the final stage of a pre-attack to ensure that the vulnerabilities of an event by security personnel are targeted, such as shift change or a period of minimum manning. In the Southwest US, Mexican cartels locate and surveil law enforcement and CBP patrols to drop their illicit payloads where they are not patrolling.

The threat of cyber involving unmanned systems is a threat arena that is a largely unexplored arena for government and corporations. UAS cyber threats, soon to migrate to USS and UUS platforms, are categorized as either external or internal. An external cyber threat occurs when a bad actor utilized a UAS platform to launch a cyberattack. The UAS can be sent to a specific location and, once there, access available networks and install malware to allow hackers to access the networks and use them remotely for nefarious purposes. An internal cyber threat involves using UAS to compromise security protocols and ongoing operations. The US government recently announced a ban on the purchase and use of DJI UAS due to concerns that the Chinese-manufactured products had software that fed all accumulated data, video, and audible recordings back to China. This resulted in a Department of Justice (DOJ) ban by adding them to the US Department of Commerce's Entity List, thereby precluding all departments and agencies from purchasing or using DJI products (Gartenberg and Brandom 2020; Mozur 2017).

It is a sad fact that more research is being performed, more in-depth analysis of capabilities and vulnerabilities of modern societies, and more papers are being written by academics, popular and social media, and bloggers than by federal government officials. Considering the size of the bureaucracy we are referring to, it should come as no surprise, plus the upper echelon of government is not comprised of individuals with doctorates, researchers, strategists, nor writers.

An IT security consultant named Sander Walters conducted extensive testing and documented his exploitation of various COTS UAS five years ago, resulting from poor passwords, default settings, and networks (Walters 2016). The exploited UAS systems ranged in price from $1,500 to $3,000, and another IT security consultant did the same on a high-end, $35,000 UAS (Walters 2016). In 2017, research by a doctoral candidate at the University of Texas at Dallas, named Junia Valente, led to "the U.S. Computer Emergency Readiness Team (US-CERT) in 2017 to issue a Vulnerability Note on a family of quadcopters that Valente demonstrated could be anonymously hijacked through their local FTP network;" other university research has

> "demonstrated the vulnerability of commercially available UAS to a wide range of attacks, including internet-facing botnet attacks, ad-hoc network attacks, data collection and probing, UAS location detection and tracking, UAS hijacking or takedown, media capture, and the modification of software to allow UAS entry of FCC prohibited airspace." (Reed, Geis, and Dietrich 2011; Ronen, Shamir, Weingarten, and O'Flynn 2017; Kerns, Shepard, Vhatti, and Hymphreys 2014; Trujano, Chan, Beams, and Rivera 2016; Valente and Cardenas 2017)

The following four vignettes describe how university researchers could launch cyberattacks or intrusions using UAS available to the public. The first two were direct attacks using UAS, and the second two used a UAS to fly close enough to a target to gather data and install malware. In the first vignette, the Massachusetts Institute of Technology (MIT) researchers exploited poor device password security to gain root access into all the UAS systems; further vulnerabilities allowed the researchers to bypass software embedded airspace restrictions (Trujano et al. 2016).

In the second vignette, researchers at the University of Texas at Austin used a spoofing attack, replacing legitimate satellite signals with counterfeit ones, thereby allowing the researchers to gain complete operational control of the UAS (Kearns et al. 2014). In a third vignette, researchers at the Stevens Institute of Technology used a single UAS to build and launch a "hidden Internet-facing botnet

(Reed et al. 2014). A botnet refers to "a network of malware-compromised devices that can be used to attack network-connected devices and they represent a major cybersecurity risk as they can be used to execute distributed-denial-of-service (DDOS) attacks, steal data, and hijack devices (Reed et al. 2014). The researchers were able to "execute DDOS attacks, steal data, and hijack devices, or in simple terms, to use commercially available drones to anonymously initiate cyberattacks via botnets" (Reed et al. 2014).

In the fourth vignette, researchers "in Israel and Canada used a commercial UAS to inject a worm and take control of the lightbulbs inside a building in Israel and made the lightbulbs blink SOS in Morse code" (Ronen et al. 2017). In this case, the use of the UAS allowed researchers to get close to their targets and launch malware into networks.

What would the purpose of an attack be?

Across the globe, there are many sporting events and other events that draw large crowds. One can only imagine the live media coverage of an attack and the aftermath. To glimpse this, one only must look at the 2017 Las Vegas shooting by sixty-four-year-old Stephen Paddock upon a crowd attending a nearby music festival (Blankstein, Williams, Elbaum, and Chuck 2017). Aside from a terrorist attack by one or more individuals, imagine the potential of modern sporting events. It is the time of year when we are confronted with a total of forty-four bowl games featuring the top eighty-eight college teams in the nation; imagine the pregame practices executed far from prying eyes (ESPN, n.d.). If that practice field was uncovered, a UAS flying overhead could conduct surveillance of an opponent and capture a great deal of "intelligence" on them"—what their formations will look like, who gets the ball more and runs to which direction more, what blockers do, who does the quarterback favor when passing, and which direction does he most favor when rolling out for a pass play.

If the field were covered, a UAS could fly inside and sit atop a beam high up in the stadium with its camera recording all the activity below. If it was fitted with an extended life battery or a solar panel that can provide extended charge to the battery, it could stay longer, all the while filming and streaming intelligence to its owner. In that case, the surveillance could be fed live to a nearby receiver that the opponent would be extremely interested in and pay top dollar for. While individual motivations will vary on why one did it, we find ourselves in a world ripe with opportunities to facilitate espionage and to receive substantive rewards in return. In many cases, our security personnel, protocols, and equipment are simply not up to the task for which their creators did not envision even a few short years prior.

Prisons and correctional facilities are among the most targeted UAS drops within the nation. Frequently, weapons, illegal drugs and other contraband are being dropped by low flying or hovering UAS to prisoners. It is a misconception that guards have prisoners under constant and strict supervision when allowed outside for fresh air and exercise. Often, the UAS will drop their packages near a point inside the facility with poor surveillance. When prisoners are released for exercise outside or perform their assigned duties, they will retrieve the small packages and hide them in their clothing until they return to their cells and relative privacy. Another misconception is that prisoners are searched every time they return from the outside to return inside. Several analysts purport that the pandemic accelerated this illegal activity as visitation was curtailed to contain the spread of COVID-19.

Critical infrastructure facilities, including airports, would be major targets of great interest. As part of the US critical infrastructure transportation sector, airports are high-value targets for bad actors and can easily be shut down. In 2018, Gatwick Airport in London had operations halted for two days as authorities were hampered by a commercial UAS that was spotted over a runway and terminal. The UAS overflight resulted in a forty-eight-hour shutdown during the Christmas holidays and racked up $65 million in costs to travelers, airlines, and the airport. The Colonial Pipeline ransomware hack demonstrated the ease and fragility of our energy delivery systems. Imagine a well-orchestrated, coordinated, and timed attack using unmanned systems, UAS, USS, and UUS, to swarm multiple designated targets to create havoc and to shut down operations. For example, targeting a series of nuclear power plants in the eastern half of the US would, if successful, create rolling blackouts. The US relies on ninety-six nuclear power plants, primarily located in the Eastern US, to provide 19.7 percent of the US daily electrical needs. In the past five years, there have been fifty-seven instances of unidentified UAS conducting overflights or harassing security personnel at US nuclear power plants. The security personnel were powerless to stop them or intervene in any meaningful manner. The group orchestrating an attack would get global visibly. An obvious advantage in utilizing unmanned systems to launch an attack, or a series of them, would be anonymity; no one would be put in a position to be killed or injured. Thus, the usual interrogation methods and forensics would be greatly reduced, and counter systems do not have a high success rate, nor are they allowed for widespread use by law enforcement and security forces within the US. The US NRC does not view UAS as a threat and has prohibited the owner/operators of nuclear power plants from procuring or using C-UAS systems at the ninety-six nuclear power plants.

The UAS with dangling ropes capped with conductive material that crashed near a Pennsylvania power substation last year was the first known intentional attack against US critical infrastructure. According to a joint intelligence memorandum by the FBI, DHS, and the National Counterterrorism Center, the July 2020 incident was the first known case of a "modified unmanned aircraft system likely being used in the U.S.to target the energy infrastructure" (Lyngaas 2021). The culprit who modified the UAS tried to create a "short circuit to cause damage to transformers or distribution lines, based on the design and recovery location" (Lyngaas 2021). The UAS "appeared to be heavily worn, indicating it was previously flown and modified for this single flight" (Lyngaas 2021). The federal government is focused on cybersecurity, but we are not paying attention to the physical threat to the grid. According to the same intelligence memorandum, the federal government fully expects UAS overflight activities to increase and expand overall US critical infrastructure (Lyngaas 2021). Virtually anyone can attach some type of explosive to a UAS and fly it into what the owner-operator wants to target. The culprit behind the Pennsylvania power station attack seems to have doomed his action by removing the UAS camera, and their line-of-sight attempt to navigate the UAS from its launch point to its intended target caused it to crash into a nearby building, effectively ending its deadly mission. UAS activity will only increase over critical infrastructure, and there will likely be more attacks.

Can we protect critical infrastructure?

US Air Force Colonel Kristopher Struve, the vice director of operations for NORAD, discussed domestic critical infrastructure defense during a virtual roundtable on air and missile defense that the Missile Defense Advocacy Association (Trevithick 2021b). He stated that our strategic competitors had increased their capabilities to launch long-range conventional strikes, including air, sea, and submarine-launched cruise missiles and new hypersonic missiles. They have now removed the notion that the US homeland is a sanctuary as it was during World War I and II, and for a brief period following the Second World War until the Soviet Union developed a nuclear inventory and long-range missiles capable of bringing them to our cities. As a result, the North American Aerospace Defense Command (NORAD) has planned to deploy ground-based air defense units to protect critical infrastructure (Trevithick 2021). This may be a public acknowledgment of the growing threat to the homeland and to our critical infrastructure, which an adversary would certainly target, but it overlooks the obvious challenges of air and missile defenses by our military. The US Patriot air and missile defense system resides in the US Army, as well as fifteen allies, and could certainly be deployed to protect a small number of critical infrastructure facilities. Colonel Struve described the situation as such:

> "Our potential adversaries have created the significant capacity to reach us asymmetrically. Our forward layers, allies, partners, forward combatant commands, and geographic commands

have largely kept those threats away from the United States. But as we look into threats from cyber actors, space threats, as well as kinetic conventional cruise missiles, which have [seen] significant improvement on the part of China and Russia in recent years, those create avenues that can create havoc in the homeland while we are trying to project our power forward to potentially a regional conflict. So, the thing that I really want to emphasize here is that the homeland is not a sanctuary any longer. There are opportunities for our adversaries to employ weapons from distances that could strike critical infrastructure in the United States early in a conflict and create some challenges for us to produce our military power. And when we look at our deterrence model, we've had a lot of capability since really World War II to have deterrence by punishment. That nuclear deterrent is the underpinning of our entire deterrence model. And it's that ability for us to respond in kind and protect our homeland. But as our adversaries have built the capability to strike us conventionally, they feel like that they have [an] opportunity below the nuclear threshold to strike us and potentially keep that conflict from going nuclear. And it is this avenue where we really need to work to close gaps and be able to protect the homeland more completely. Something that I haven't really talked about is what are we doing on the risk mitigation front, and that's the ability for us to actually defend. And there are two sides that we think about when we think of risk mitigation. It's our ability to deny those threats, and, when we talk about these kinetic conventional weapons, it's hardening, redundancy, resiliency on some of our critical infrastructure. Not every piece of our infrastructure is going to be taken out with one of these conventional-size weapons, but some things are particularly vulnerable. So, we view this as a whole-of-government approach to being able to protect that infrastructure. And then, after that, we would place defenses on key critical infrastructure nodes. It's infeasible that we can place kinetic, you know, surface-to-air missile batteries over the entirety of the United States, Alaska, the Aleutian Islands, and Canada. By the time we fielded such a system, they would've found a way around it anyway. So, we are working closely with the Office of the Secretary of Defense and the National Security Council on where we are going to place those limited ground-based air defense assets that we'll have in time of conflict to really change our adversaries' calculus about their efficacy of being able to execute an attack on the U.S. That can be everything from how we're organized today, which would be our fighter aircraft at more than a dozen, a couple dozen locations across the U.S., which can intercept conventional cruise missiles with mixes of AESA and non-AESA fighters, as well as Alaska and Canada. It's our limited area defenses, you know Patriot-type missile systems that we can deploy in time of crisis. But it's also our persistent capability that we currently employ in specific areas, such as the National Capital Region. Those are systems that, based on increasing threats from both the air and sea, that we need to be able to continue to develop, generate and put in those key critical infrastructure locations so that we can change the calculus of our adversaries." (Trevithick 2021)

A solution from our past that DOD and DHS have discussed would be to rebuild a nationwide network of surface-to-air missile defense installations. This endeavor would be expensive: a single Patriot battery with a basic load of interceptor missiles costs $100 million, and to procure enough from its manufacturer Raytheon, for nationwide coverage of key cities, critical infrastructure, and high-value targets such as the White House, would take decades to build. In total, the US Army possesses eighteen Patriot air defense battalions, each equipped with five subordinate units, referred to as batteries. Assuming each battery protected a single critical infrastructure facility, the army coverage would protect at most ninety sites, far less than the twenty thousand critical infrastructure facilities in the US. It is also important to note that five of those battalions are permanently stationed overseas to support Allied nations. Two are training units. The remainder has on-call missions to deploy overseas in support of combat operations. Our two strategic competitors, Russia and China, have surface-to-air missile (SAM) sites in their countries protecting against the threat of key targets, but with any military system, offensive or defensive, there are always countermeasures that can be fielded to negate them. The US did possess a large SAM network through much of the Cold War period but canceled them in the 1970s due to the expense of equipment, manning, training, missiles, operating costs; in addition, the US Air Force stated that they would be able to launch fighter aircraft and respond quickly from multiple airbases located throughout the country to engage any potential air threat so there was no need for SAM sites in the continental US. At present, the military forces of the US do not have the inventory of SAM units to deploy throughout the nation and protect even a small minority of critical infrastructure facilities. The US does have a National Advanced Surface to Air Missile System (NASAMS) with multiple missiles and guns systems, supported by radars and other sensor systems to detect multiple types of threats, protecting the National Capital Region (NCR), which was a result of the 9/11 attacks. In October 2020, a Patriot battalion stationed at Fort Hood, Texas, "deployed" 107 miles to the Easterwood Airport in College Station, Texas, emplaced the system, and demonstrated that if called upon, the army system could relocate to a critical infrastructure facility and provide air and missile defense protection.

The army purchased two Israeli-made Iron Dome batteries, counter rocket, artillery, mortar fire, and a cruise missile system. This system would help provide point defense at a far lower cost than Patriot of US critical infrastructure facilities; unfortunately, one is in Guam, and the other is in Texas on standby for future missions. The US Army claims to be developing a homegrown system to replace the Israeli Iron Dome, but in the meantime, at least one of those two batteries is in Guam protecting against Chinese aggression should Guam be needed as a staging platform for the US military in the Asian theater. The US purchased two batteries of the Israeli Iron Dome air defense system, yet these missiles cost $40,000 each, and the offensive missiles and UAS that they are intercepting are $1,000 to $10,000 (Majumdar 2017; Weisgerber 2017). Israel developed Iron Dome to counter the onslaught of unguided rockets and SCUD missiles that the Iranian-supported Palestinian militants fire indiscriminately into Israel (Judson 2019). All modern-day missile defense systems, such as the US Patriot, THAAD, or the Ground-based Midcourse Defense (GMD) antiballistic missile system in Alaska, the Russian S-400, Israel's Iron Dome, are all far more expensive than the missiles they intercept (Roblin 2017).

The US Army and the US Marine Corps have dedicated short-range air defense (SHORAD) units equipped with the Avenger system, a vehicle-mounted air defense system equipped with eight ready-to-fire Stinger missiles and a .50-caliber machine gun. This is a short-range point-defense system, but it may be getting a major upgrade. Currently, the Stinger missile has an active seeker in the nose that locks onto the heat source from aircraft engines. The manufacturer, Raytheon, has developed prototypes equipped with a fragmentary warhead that will detonate when it detects that it is near a target and the detonation will direct a large cloud of steel pieces toward the target. This would provide a greater probability of intercepting mortar rounds, maneuvering cruise missiles, and UAS. The Stinger missile has accumulated an impressive war record; it is reported to have downed more than 270 fixed-wing and rotary aircraft in four major conflicts and can be fired while handheld or from vehicles and aircraft (Majumdar 2017). The Stinger gained notoriety for its use by the Mujahideen during the Soviet occupation of Afghanistan in the 1980s. The new Stinger is designed to counter the growing threat posed by UAS on the modern battlefield and costs roughly $40,000, thus the cost-exchange ratio is still on the side of the enemy as most UAS being used on modern battlefields are readily available commercial ones that cost far less than $10,000 (Majumdar 2017).

The US Army is beginning to field a new SHORAD weapons system on an eight-wheeled lightly armored vehicle which has four Stinger missiles, two Hellfire missiles that are tank killers, a 7.62-millimeter machine gun, a 30-millimeter automatic cannon, an electronic warfare package to jam UAS control links without shooting them, and a multi-mission radar to track both air and ground targets. Military SHORAD systems, which could be extraordinarily successful against UAS and other threats attacking critical infrastructure, are

Image courtesy of Breaking Defense, https://breakingdefense.com/2018/07/army-anti-aircraft-stryker-can-kill-tanks-too/

simply too few to be truly effective. Also, it would be highly unlikely unless there was intelligence of weapons of mass destruction (WMD) payloads that the federal government would allow kinetic weapons and missiles to be fired up into US airspace. There is potential for an errant hit on privately owned or commercial aircraft, or just as unsavory, the bullets and missiles that are to be fired into the sky would eventually fall back to earth and could cause considerable damage to citizens, businesses, and metropolitan areas. Even if an armed conflict were to occur, whether a terrorist attack or a declaration of war, air defense of critical infrastructure would not be possible without a highly integrated sensor network to direct engagements if the nation's leadership authorized the engagement by air defense units.

NORAD has diligently developed improved distributed sensor and networking capabilities and interception capabilities as part of the Strategic Homeland Integrated Ecosystems for Layered Defense (SHIELD) to counter the growing threat to the homeland. The fastest-growing threat to critical infrastructure facilities is COTS UAS and low-flying cruise missiles. The Chinese have flooded global markets, and purchasing these systems has become far easier, and the demand for low-cost systems to counter this threat has never been greater (Roblin 2019b). One aspect being improved on existing aircraft is the radars on the US Air Force F-16 fighters, which are stationed throughout the nation at multiple airbases in all fifty states. The radar improvements would give the fighters the ability to detect and engage threats, including low-flying UAS and cruise missiles (Hunter 2020). Other improvements would be the addition of new air-to-air missiles, streamlining the Patriot advanced engagement control system and missile launchers to create a less expensive nonmobile version that would serve as a stationary air defense platform that could be positioned near major critical infrastructure facilities. Even if the new systems are funded and begin production and emplacement around the nation, there will be facilities that receive coverage and many more that will not work due to the limited number of surface-to-air missile systems. In a real world crisis, either from a terrorist group or one resourced by our strategic competitors, we will have a crisis when critical infrastructure facilities are targeted and hit due to their obvious vulnerabilities to attack.

After many years of being shortsighted, the US Army has realized that it has a critical vulnerability: "deploying more SHORAD capabilities is now one of its six top modernization priorities" (Roblin 2018). In addition, the army "is spending billions to improve its existing Patriot and THAAD systems by tying together dispersed radars and fire-control systems into an Integrated Air and Missile Defense Battle Command System (IBCS) network" (Roblin 2018). Thus, number five on the army's top six list to defeat Russia and China in a war is air and missile defense (Roblin 2019c).

How did we get into this predicament?

Simply put our government ignored the threat posed by our unmanned systems that originated with the first crude, mechanically controlled modern-day UAS, made its appearance. The Kettering Bug was developed toward the end of World War I but never saw action. The use of UAS by the US grew slightly in World War II and grew in complexity, use, and numbers in the Vietnam War and were demonstrated to the world during the Balkan conflict. The US developed an offensive UAS capability that amazed and befuddled out enemies, yet it failed miserably to develop counter systems and strategies in case our enemies developed offensive UAS systems as well. Frankly, the defensive aspect was ignored until the last decade and has only gotten serious in the last few years.

On November 9, 1989, I was a 1LT in the army and the regimental air defense officer (RADO) in the Eleventh Armored Cavalry Regiment stationed in Fulda, Germany. I was driving into work at 6:00 AM when I was surprised by the sight of East German Trabant automobiles, many with wide-eyed children staring out in seeming disbelief. I floored my Saab 900 Turbo and sped to work, wondering what had occurred while I had slept. Upon reaching my headquarters building located atop the Fourth Squadron Aviation airfield, I was promptly informed that the inter-German wall had come down and East Germans were streaming across to the West. Two years later, on December 26, 1991, the Union of the Soviet Socialist Republics (USSR) was formally dissolved due to "General Secretary Mikhail Gorbachev's effort of political and economic reformation of the Soviet authoritarian system" (Clines 1991). With the demise of the number one stated enemy of the US, it in turn significantly downsized its standing military force, and its SHORAD systems were decimated in the short-sighted, misguided belief that they were no longer needed (Rogoway 2017).

Russia "has taken the SHORAD mission far more seriously than the U.S., namely because it has not had the luxury of nearly guaranteed air supremacy, or perhaps it never believed in that idea in the first place" (Grau and Bartles 2016). In *The Russian Way of War*, Grau and Bartles detail the

> "multi-layered air defense systems that accompany Russian army battalion tactical groups into battle. Russia has fielded the most modern integrated ground-based tactical air-defense system on the planet. Every brigade, each with up to four 900-person battalion tactical groups, travels with an air-defense battalion." (Grau and Bartles 2016)

Today, multiple Russian-built mobile point air defense systems are in mass production and being exported to Russian supporters worldwide. Foremost among these systems is the highly formidable Pantsir-S1 (pictured below). As mentioned previously in *Unmanned Systems: Savior or Threat*, a pro-US group in Libya overran a pro-Russia group and captured a Pantsir intact. A US Air Force C-17 landed soon after loading and transporting the captured Russia SHORAD system back to Wright Patterson AFB in Ohio for analysis and testing.

Image courtesy of The WarZone, Caution-https://www.thedrive.com/the-war-zone/13284/ americas-gaping-short-range-air-defense-gap-and-why-it-has-to-be-closed-immediately

The successful attack on Saudi critical infrastructure

The 2019 attack by UAS and cruise missiles on the Saudi Arabia oil field and refinery at Abqaiq demonstrated to the world that their critical infrastructure was highly vulnerable to cruise missile and UAS attack. The US had intentionally focused much of its air defense funding for improvements to the Patriot system with little going toward the threat that SHORAD would be ideal for. Up to the attack in France, virtually no one envisioned a threat from unmanned systems, certainly not on the modern battlefield nor against critical infrastructure within a nation. The US's focus was on the threat posed by ballistic missiles that fly from launch to target in a long arc. As a result of US investment in Patriot and Raytheon and Lockheed Martin's success in the engineering arena, the Saudi Patriot units have indeed intercepted hundreds of cheap antiquated ballistic missiles from the Houthi rebels in Yemen rebels in the past five to ten years (Opall-Rome 2017). In 2014, the French nuclear plant near Lyons was "attacked" by two large UAS. Both flew toward one of the reactor buildings when one stopped suddenly, apparently to film the attack. The other continued and flew into the building, where it smashed itself into small pieces.

Prior to this event, few gave serious thought to the intentional use of UAS to attack a nation's critical infrastructure. For an air defense planner, the Saudi oil facilities were located under a Patriot defensive umbrella which was capable of intercepting aircraft and missiles up to one hundred miles away (Roblin 2019a). It is unknown if the Patriot units were operational and if they were manned at the time that the cruise missiles carrying a dozen or more UAS flew near the oilfield and refinery. Once the Saudi personnel realized that they were under attack, they fired small arms into the air which may have helped damage the facilities as the bullets that went straight up came straight down (Kalin and Westall 2019). The attack resulted in a "disruption of shipments of 5.7 million barrels of oil daily, half of Saudi Arabia's normal daily total shipments, and five percent of the global supply of oil" (Reuters 2019). Post attack forensics determined that the cruise missiles used were based on cheap fifty-year-old Russian designs. (Qiblawi 2019; Lister 2019).

Saudi had command and control of their air defense SHORAD, and Patriot units split between the Ministry of the Interior and the Ministry of Defense. Based on my nearly thirty years in the US Army Air Defense Artillery branch, I can attest to the fact that this was a colossal organizational mistake ripe with failure. UAS and cruise missiles typically fly close to the ground to avoid being seen by radars and other sensors, which reduces the detection and reaction time by trained air defense crews. UAS and cruise missiles possess significant maneuverability, which is why they would be programmed to fly between Patriot units and their radar coverage. The Saudi government has reported that the Houthi rebels have launched 1,204 UAS and 1,207 ballistic missiles and rockets into Saudi in the past five years. As a result, the Saudi Arabian supply of Patriot missiles that cost between $3 and $6 million each, depending on the model of the missiles, is critically low. Due to the Yemen civil war, Saudi Arabia has altered the focus of its air defense units away from Iran and toward Yemen (Kalin and Westall 2017). The need for defense in depth of multiple SHORAD units within the visual range of defended facilities is now important. Some inbound threats, due to their flight characteristics, will not be detected by SAM systems until they are relatively close to the targeted facilities. Firing long-range and expensive Patriot missiles against UAS that cost $10,000 or less cannot be maintained indefinitely, even for Saudi Arabia. ISIS operators in Syria managed to use COTS UAS for surveillance and combat operations based on the inexpensive yet highly capable camera that came installed with the DJI commercial UAS.

It has been speculated that Saudi Arabia is interested in the Chinese Silent Hunter I and II C-UAS laser system and Russian Pantsir system to supplement their existing SHORAD systems, but the Saudi government has officially denied interest in the Chinese systems and Russian systems. Not even an oil-rich Gulf state can continue to expend $3 to $6 million Patriot missiles against inbound Houthi rebel UAS that cost, at most, $10,000 to $20,000 each. China has become emboldened recently, and their ruling party has decided that they are ready militarily and economically to make their move upon the global leader in both areas. It appears that President Xi is intent on solidifying his legacy, namely that he led China out of its regional status and turned it into a global powerhouse. China has concentrated on building commercial ties with countries across the globe, even if it means augmenting their industries until such time that they have driven the competition to ruin. They are masters at taking a vision to market faster than any other country. This flexing of its newfound confidence and muscle has translated into new and vastly improved military capabilities to reinforce its forays into areas of the globe where it previously did not have a foothold. China is exploiting the reluctance of the US to sell its military UAS to key Middle East partners. Unless this reluctance is stopped quickly, the US will lose major influence in the region to China in the UAS market and other weapon sales.

This, combined with a repeat of the Vietnam withdrawal out of Afghanistan, has allies and partner nations across the globe wondering if they can depend on the US. Our reluctance to sell front-line military equipment has undermined US leverage, interest, and endangered fundamental US national security interests. This may help explain why the Biden administration recently cleared the Saudi

Arabia purchase of $650 million in air defense replacement missiles this week, expended in its ongoing war with Yemen-based Houthi rebels (Zengerle 2021). According to press reports, this administration also approved a plan to move forward with "$23 billion in weapons sales to the United Arab Emirates, which include the F-35 aircraft, MQ-9B armed drones, and other equipment" (Bowman, Thompson, and Brobst 2021; Zengerle 2021).

UAS mother ships

A recent new development in the UAS world has been the development of a UAS mother ship capable of deploying multiple smaller UAS. This could be an alternate and far cheaper method to protect critical infrastructure than the emplacement of air defense units or equipment around the nation. The rapid deployment or placement of a UAS mothership in the center of a node of nearby critical infrastructure facilities would facilitate defensive capabilities. It could launch and deploy smaller UAS when the target(s) approached one of those facilities, all at a far lower cost than SAM units and manned fighter aircraft. This is a far more refined system than the primitive though effective use of cruise missile mother ships releasing explosive laden smaller UAS to attack Saudi Arabia's Abqaiq refinery in 2019.

The Chinese are testing a true UAS mothership capable of releasing a UAS swarm for reconnaissance or loaded with explosives for an attack (Osborn 2021). In this case, the Chinese UAS mother ship is a fixed-wing, vertical takeoff and landing system that carried nine smaller UAS in its debut (Xuanzun 2021). This new system was manufactured by a China-based company, Zhongtian Feilong, and has been engineered for long-range missions with options for different payloads and missions and anti-jamming capabilities (Xuanzun 2021). Observers reported that the mother ship flew to its target during the test, opened its payload compartment, and released sUAS to create a swarm (Xuanzun 2021). Such a system would be ideal for highly congested areas where one would not wish to put pilots in danger or risk the more expensive manned aircraft. Launching an inexpensive UAS mother ship to autonomously fly to its designated target, release a load of surveillance or explosive-equipped sUAS, then fly home to rearm and complete another mission in short order would alter the equation of frontline combat.

Consider the implications of an ability to launch multiple mother ships, each carrying nine or more sUAS, to swarm and overwhelm a designated targets defense or multiple targets simultaneously. Airfields would not be needed near the conflict area as ground soldiers could launch UAS motherships from the rear of their vehicles or carry them into contact with the enemy in backpacks. Thus, the shortened reaction times, the endurance of such systems, their increased operational range for frontline soldiers, and survivability would be a true combat multiplier on the modern-day battlefield. The sUAS could conduct surveillance, test enemy defenses and air defense capabilities, or simply conduct attacks at all hours from multiple points of attack, thereby depriving enemy forces of rest. One of the biggest fears of field commanders is a sleep-deprived force whose effectiveness will suffer and worsen to the point that they will become a combat-ineffective force. If the sUAS are networked, then there will be an inherently high degree of redundancy if one or more are destroyed; others will continue with their assigned mission.

The US Air Force has been working with the Defense Advanced Research Projects Agency (DARPA) on a UAS mother ship concept that is a well step beyond the Chinese. The Gremlin mothership system is a large cruise missile type flying device that will release sUAS over one or more targeted areas, then fly to a rendezvous point currently with a C-130 cargo aircraft, but other aircraft could easily be used as well to retrieve it and pull it into the aircraft for rearming, refueling, and another mission. The Gremlin system will release a semiautonomous or autonomous swarm capable of conducting multiple missions; they could conduct intelligence, surveillance, reconnaissance, launch of electronic attacks, kinetic attacks using explosives, or to overwhelm enemy air defenses (Trevithick 2019). One or more Gremlins could be launched to attack multiple targets by launching multiple UAS swarms to converge on a target. The Gremlin could be readily altered for different payloads of up to 150 pounds, networked with other autonomous systems that would allow it and other Gremlins to instantly swarm a target depending on the threat in the environment in which it is operating (Rogoway 2018). The Gremlins are projected to be relatively inexpensive and could be used for missions deemed too dangerous for manned aircraft. The Gremlin could also deploy quickly from US Air Force bases around the nation to serve as C-UAS or even counter cruise missile systems ready to deploy over an area, including multiple critical infrastructure facilities, or move from facility to facility ready to engage inbound threats. UAS could overwhelm defenses with their superior numbers.

C-UAS laser systems

One of the problems that have emerged recently, primarily in war zones applicable to protecting critical infrastructure worldwide, is how one could effectively counter the threat posed by swarms of UAS. The threat posed by swarms and possibly smaller grouped UAS

Image courtesy of https://www.darpa.mil/news-events/2021-11-05

that are both expendable and deadly in their payloads can easily overwhelm defenses and attack their targets. Today's missiles designed for aircraft, cruise missiles, or ballistic missiles are simply too expensive for use against swarms of small attackers. Welcome C-UAS laser systems, which offer exceptionally low cost per shot, are reusable and non-kinetic, offer tremendous precision in targeting the beam upon a target, and cannot be jammed. Currently, China, Russia, and the US are major investors pursuing laser systems. If focused shot is extremely inexpensive. If a laser system is connected to a steady power source, it could fire indefinitely. Lasers are highly accurate and offer fast reaction times; if they are focused long enough on inbound targets, they can and will burn a hole into UAS, missiles, aircraft, or ground systems. The development costs of these systems have been expensive, but the cost per engagement of targets is quite low even if fired against a single target or in rapid succession against multiple targets. The beams of concentrated light are as fast; in the vacuum of outer space, light travels at 186,282 miles per second; in earth's atmosphere, it can be slowed down slightly due to refraction (Stein 2021; Yang 2018). The precision offered by lasers could also be used to specifically target a vehicle's engine without killing the human occupants. Another benefit is that lasers do not produce sound, they are invisible to the human eye, and do not create smoke or flame when fired, which makes them highly stealthy.

For many years, the development of laser systems, on ground-based vehicles, aircraft, and ships have been slowed due to several factors including beam diffusion, beam weakening when sand, smoke, or fog are in its path. Beams may also require several seconds of focused precision on a specific area to inflict damage. The US Army has successfully used a laser system at White Sands, New Mexico, to down rocket and mortar shells (Hamilton 2009). The two issues in fielding the system to the force were making the system compact enough to fit in bed of a vehicle and equipping the vehicle with an electrical power plant that could provide the power to the systems for multiple or continuous engagements on the battlefield. In a recent unveiling, the army apparently solved both issues with its new High Energy Laser Mobile Demonstrator (HEMD) (Skilling 2013).

Image courtesy of the Boeing, https://secure.boeingimages.com/archive/Boeing-High-Energy-Laser-Mobile-Demonstrator--HEL-MD--2F3XC5PWELX.html

The army is also "developing a vehicle-mounted 100-kilowatt laser that could be used to cost-efficiently burn drones out of the sky" (Roblin 2017). In 2017, the US Army used a helicopter for the first time to successfully demonstrate firing a laser beam from a pod mounted on a helicopter against multiple targets at White Sands in New Mexico (Roblin 2017). The Israeli Ministry of Defense recently announced that it was taking this technology to build a protective "laser wall" across portions of its border to defend against missile, rocket, UAS, and other projectile attacks (Israeli Ministry of Defense 2022).

Recently, a squad of US Marines used a vehicle-mounted mobile jammer that was strapped to the deck of a naval vessel in the Red Sea carrier to bring down an Iranian drone deemed to be too close with possible hostile intent (Mizokami 2019). The US and Russia have been working separately but intently on the "development of electronic warfare systems that can disrupt or hijack the communications links between UAS and their operators" (Roblin 2017). Another step to building low-cost C-UAS systems that can protect critical infrastructure is to add AI. By working under intelligent capabilities that can think for itself to the point of accomplishing its programmed mission, the C-UAS system Anvil can sit in its charger, awaiting a sensor detection of an inbound threat to launch to a specific coordinate by its AI master. Once deployed, Anvil can perform multiple intercept missions until it needs a recharge. AI would already have sent another Anvil, or multiple Anvils, to replace the first one. This is one way that the US could capitalize on its industrial base and technological prowess by joining a C-UAS with AI management.

Multiple facilities could be protected against UAS attacks at a fraction of the cost of purchasing a nonmobile Patriot system. Two main factors are holding the US up in executing smartly: the first is the perceived need to have a human in the loop. Only in the US do the human masters believe that only a human should decide when to engage in a threat when it may involve taking a human life. Yet this was not a concern to a relative newcomer to the world of manufacturing and exporting UAS Turkey. Since Libyan ruler Colonel Qadhafi was killed in an uprising against his decades of repressive rule, there have been two subsequent civil wars. During the second one, the interim government attacked its primary rival using Turkish-made Kargu-2 UAS. This engagement is especially important because it marks the first time that autonomous AI-enhanced UAS were programmed with the autonomy to determine for themselves what constituted a viable target and to engage it even if human beings were in the targeted areas, all without human supervision or authorization to proceed (Mizokami 2021). According to the United Nations investigators and subsequent reports, the UAS was "programmed to attack targets without requiring data connectivity between the operator and munition" (United Nations Security Council 2021).

The second US hesitation is tied to the first and is the concept of morality. This self-imposed hesitation is akin to the famous charge of the Polish horsemen at the outbreak of World War II against the tanks, artillery, and machine-gun-equipped German Nazi army. The Polish force was attempting to defend its border against the approaching Germans. When the Germans saw the force galloping toward it, the German commander faced no moral quandary and quickly ordered his forces to fire upon the horsemen. In a matter of seconds, the nineteenth-century concept of warfare was brutally annihilated and replaced by the German's twenty-first-century concept of modern warfare, one which included a combined army with armored vehicles, artillery, infantry, close air support by combat aircraft, and a slew of other combat multipliers. For China and Russia, morality does not exist and only hinders the development, deployment, or use of AI-enhanced weapon systems.

One additional advancement by a Turkish company could add an offensive, hunter-killer C-UAS system to governments and to those who purchase it, say the owner-operators of US critical infrastructure. The Eren, pictured below, is the world's first laser-armed UAS that is designed to fire at and neutralize explosives from a range of 328 to 1,640 feet (Peck 2021; Montenegro 2021). Eren hovered and was able to keep a laser beam focused on a specific point on a steel plate three millimeters thick and burned a hole through it in ninety seconds (Peck 2021; Montenegro 2021). The apparent shortfall would be that of battery life; there is no information on how many laser shots the battery pack is capable of before it must fly back to its home station for a replacement battery or recharge or if it is a single battery that is shared between the laser and the UAS.

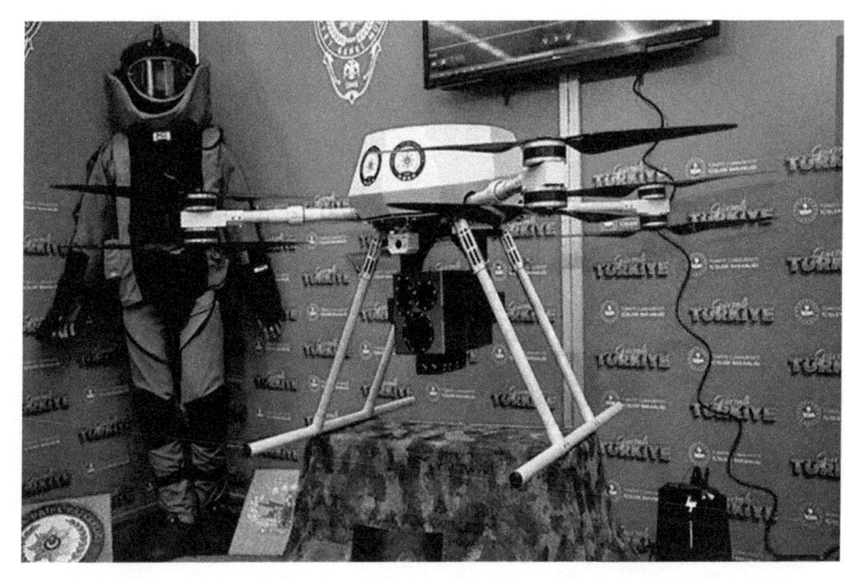

Image courtesy of the Anadolu AgencyNew Flow System, https://www.-aa-com-tr.translate.goog/tr/bilim-teknoloji/dunyanin-ilk-lazer-silahli-dronu-eren-testlerdeki-basarili-atislariyla-goz-dolduruyor/2444041?_x_tr_sl=tr&_x_tr_ tl=en&_x_tr_hl=en-US

A cooperative approach

Another approach combined the human being in the cockpit of an aircraft with an AI-enhanced UAS aircraft that flies alongside and collaborates on targets and self-defense with its human lead. Highly knowledgeable individuals have authored multiple articles on the future of airpower consisting of both manned and unmanned systems (Goure 2020). Interestingly, Elon Musk has stated that "the fighter jet era has passed" and that the future belongs to "a drone fighter plane that's remote-controlled by a human, but with its maneuvers augmented by autonomy" (Goure 2020).

Future air forces may use stealth fighters and bombers accompanied by unmanned aircraft to attack enemy air defenses to quickly achieve air superiority to enable follow-on aircraft to attack ground targets. For the defense of the homeland, it is conceivable that the US Air Force could launch fighters with unmanned aircraft alongside, in what has been referred to as loyal wingman, to engage multiple UAS, cruise missiles, or other aircraft far quicker. While the manned aircraft could investigate a suspicious aircraft, the unmanned aircraft, such as the XQ-58A Valkyrie, could engage confirmed enemy UAS, aircraft, or any other targets fed to it by the nearby manned aircraft or air controllers (Rogoway 2019). The US Navy recently unveiled its operational unmanned and fully autonomous refueling aircraft, the MQ-25 Stingray. Three primary types of unmanned wingmen are distinguished based on cost and capabilities. The first type was built to be low-cost and expendable as jammers and sensors. The second is much larger and more capable, designed to operate alongside the manned fighters to extend the operational reach of the team, each with similar capabilities. The third is the most expensive due to its high degree of sophistication. This type would conduct ISR and serve as electronic attack platforms or communications nodes and aerial refuelers.

USS

Globally, USS (atop water) manufacturers are focusing their efforts on instilling technologies that will make USS fully functional across the spectrum of military operations. These technologies included AI for enhanced autonomy, communications, navigation, sensors, customizable payloads, and highly efficient energy technologies to allow for extended operations with no dependency on human handlers. Most USS in worldwide use have a hull length less than forty-two feet long. This size is optimal for launching USS from larger ships at sea or shore. The reduction in their size to less than twenty-one feet and even smaller is due to the multitude of missions that developers envision for their craft. Nearly 90 percent of the present-day global efforts for USS are focused on small USS. The US, Russia, China, Iran, France, Norway, Singapore, Republic of Korea, United Kingdom, and the United Arab Emirates are all investing heavily into research, design, and production of USS. The primary missions initially would be in the areas of the gathering of ISR, mine countermeasures, anti-surface warfare, and anti-submarine warfare.

For those nations with small navies and limited budgets, USS platforms afford great capabilities that are modular in design with highly customizable capabilities based on mission requirements, low cost compared to costly manned ships, greatly reduced personnel and training requirements, and risk. Of existing countries, Iran has made considerable progress in USS technologies, primarily to engage superior US naval forces in the Persian Gulf, but China is investing heavily in UAS, USS, and UUS systems and will undoubtedly surpass Iranian USS systems within the next five years. Many preexisting modern naval capabilities for detecting, tracking, identifying, and engaging enemy surface ships can be adapted for use in counter USS (C-USS) threat platforms. Ports, harbors, and waterside facilities are not sanctuaries for combatant and commercial ships loading or unloading cargo from factories, facilities, or refineries producing products from critical infrastructure sectors. A USS providing security could have helped the USS Cole on October 12, 2000, when the guided-missile destroyer was heavily damaged in a small boat attack loaded with explosives by an Al-Qaeda suicide bomber against the moored USS Cole as it was being refueled in Yemen's Aden harbor (Combs and Slann 2009).

In the world of AI-enhanced autonomous USS, "Marine AI has partnered with Water Witch, a developer of marine litter collection workboats, to provide its Guardian autonomous software solution for a new generation of autonomous vessels" (Ball 2022). Water Witch has sold two hundred of its innovative vessels worldwide to remove manmade garbage from marine environments. Their newest models are AI-enhanced vessels, thus will be fully autonomous and, thanks to thrusters, will have "increased versatility, maneuverability, and portability whilst reducing operating costs" (Ball 2022).

Today's naval forces, including commercial shipping vessels, are susceptible to persistent ISR both in harbors, ports, and while operating in the open oceans. Presently there is an unacceptably poor standoff range between close-in combat systems and small fast speedboats that are conducting attacks against the far larger ships. USS is the answer to this problem. They could be deployed quickly even when underway at sea and could be armed with a warhead capable of stopping an attack by speedboats, but they could also provide ISR while underway. When networked with UUS, the picture available to the commander of the vessel would be a significantly enhanced security awareness. The ship's commander would have access to a 360-degree view of what is always around the vessel, both in the air, provided they have air defense sensors available and UAS in the air, and on top of the water, and finally, what might be lurking under the water.

The threat today is from USS and UUS that will swarm to a target. While the individual warheads might not be significant, dozens of them aiming for the same target on the ship's side or underneath at a point on its belly would certainly succeed in slowing or stopping a vessel in motion. The best way to defeat an offensive swarm may prove to be what I have proposed for years, with a defensive swarm. Vessels in transit through geographical chokepoints are highly vulnerable and would benefit from USS and UUS systems operating around and underneath it. It is likely that there is not a single solution that will address all the USS and UUS threats to naval vessels across all operational areas. It is equally likely that effective counter systems to both will require a layered defense approach using multiple sensors and engagement systems configuration to specific threat platforms and environments.

The generation of USS (on land) is only beginning to blossom. Everyone in the world is aware of what a Tesla car is, less so on electric trucks and even less on large earth-moving hybrid vehicles. Yet they exist today and are being used daily as autonomous earth-moving vehicles and trains in Australia and Canada in the mining and transport sector (Southwell 2021; Kwan 2019). The threat of an explosive-laden USS crashing through the gates of multiple sectors of critical infrastructure or into the conveyances of products of those sectors is a growing threat that private companies and government agencies have yet to address. For the nuclear energy sector, the threat to nuclear reactors and dams might not be as viable as the threat to the power lines leading away from those facilities providing electricity for millions of citizens. For multiple sectors including the chemical one, there are simply too many facilities to attack, but attacking the significantly less secure means of conveyance of those manufactured products would be far easier and enticing, especially if media coverage would be practically guaranteed.

UUS

The last area of unmanned systems undergoing tremendous growth is that is UUS. Presently, many countries are investing significant resources into the development of UUS; the numbers of UUS and the capabilities and the mission that they will be capable of performing will continue to increase and blossom as interested parties and their investments increase. In research and development and proliferation of UUS-related technologies. Currently, the development of UUS is focused on its use in search and rescue missions, undersea surveys of the seabed, minelaying and mine countermeasures, and ISR missions. In the next five to ten years, UUS will be AI-enabled and have a greater degree of autonomy that will allow them to engage surface and undersea vessels and serve as weapons themselves, such as WMDs, or as weapon delivery systems. Improvements in autonomy, AI, underwater communications networks, and the fusing of multiple sensors will only expand future roles for UUS. Let us hope that this time, unlike the threat posed by UAS today even though those systems emerged decades ago, C-UUS technological solutions will be developed simultaneously to the continued development of the UUS platforms, and not years or decades later.

In a venture sponsored by DARPA and the US Navy, the Manta Ray is an effort to demonstrate innovative technologies with AI that would allow multiple payloads to be loaded and deployed on a UUS for "long-duration, long-range missions in ocean environments" (Sakharkar 2021). The program is intended to minimize the need for humans to provide logistics or maintenance. The Manta Ray could operate at the very bottom of our oceans and be equipped with specialized payloads for specific missions, such as mapping the floor of the ocean. The Manta Ray could also deploy smaller autonomous payloads with separate missions:

Image is an artist rendition of the Manta Ray, https://www.inceptivemind.com/darpa-awards-phase-2-contracts-manta-ray-underwater-drone/22587/

Then once they are complete, return to the mothership ship (Sakharkar 2021). A system like the Manta Ray, or another one similar in capabilities, could be used to protect the US waterways, especially those upon which critical infrastructure facilities are located. Many facilities have pipelines that pull in water for cooling pumps or to expel waste products which could be used by UUS for surveillance or attack.

Whether on land, atop the water, or undersea, the growing capabilities of unmanned systems are a present-day threat to critical infrastructure. Unmanned systems can be used for short- or long-term surveillance, for delivering explosives, chemicals, biological, or radiological agents, though thankfully this has not yet occurred, to serve as movable mines, or torpedoes to attack any manner of fixed sites or ships. UUS can serve as mother ships and launch separate USS or even UAS to continue to surveil potential targets. Humans are only now venturing into a new realm of AI-enhanced unmanned systems for scientific, exploratory, commercial, and military applications.

Chapter 21
Conclusions

The vice-chairman of the Joint Chiefs of Staff (VCJCS), General John Hyten, in his farewell address, publicly stated that "China has performed hundreds of tests of hypersonic weapons in the last five years, compared to nine that the U.S. performed" (Martin, 2021). He added that a fear of failure now dominates the US military, yet it does not exist in our strategic competitors like China and the Democratic People's Republic of Korea (DPRK).

> "As opposed to his father and his grandfather, Kim Jong Un, decided not to kill the scientists and engineers when they failed. He decided to encourage and let them learn by failing and they did. And so, the 118th biggest economy in the world—118th!—has built an ICBM nuclear capability because they test and they understand risk." (Martin, 2021)

General Hyten was offering one explanation for why Russia, with a stated defense budget of 3.1 trillion rubles, the equivalent of $65 billion, has created hypersonic systems and is close to debuting the highly anticipated counter-hypersonic, counter-stealth fighters, counter-LEO (low earth orbit) satellites, S-500 IAMD system, and state-of-the-art C-UAS systems well ahead of the US efforts with similar systems with its gargantuan $753.5 billion annual defense budget (Maucione, 2021; Kofman & Connolly, 2019). The DPRK claims to be testing hypersonic ballistic and cruise missiles (Tian, 2021; Martin, 2021). If true, these two systems alone will significantly impact the ability of the US to provide point and area defense of key metropolitan and military bases in the Republic of Korea (ROK). If one considers the use of market exchange rates, which significantly understates the true state of military expenditure for smaller per-capita incomes that exist in Russia, China, and the DPRK, the actual expenditure is much higher (Kofman & Connolly, 2019). A comparative expenditure of defense budgets should instead be based on purchasing power parity (PPP) exchange rates rather than market exchange rates as this method takes into account the differences in costs within different countries. Through PPP analysis, Russia has spent three times more than its stated budget for defense, or $200 billion; and China, which has a stated defense budget of 1.27 trillion yuan, is closer to $252 billion (Funaiole & Hart, 2021). One other point that needs to be addressed is that of change. If the second most senior general in the DOD complains about our nonefficient acquisition, testing, and procurement processes, the obvious question is this: What did he do to fix it? What did he do to get the DOD acquisition efforts on the right path? As the VCJCS, there was plenty that he had responsibility for referencing the DOD acquisition process and much that he could have corrected. The Defense Reorganization Act of 1986 (Goldwater-Nichols legislation) provided the statutory basis for the chairman of the Joint Chiefs of Staff (CJCS) to review major personnel, matériel, and logistics requirements of the Armed Services in relation to plans, programs, and budgets. The chairman has delegated the VCJCS as the chairman of the Joint Requirements Oversight Committee (JROC) to fulfill the chairman's responsibilities outlined in Title 10 of the United States Code to provide advice to the secretary of defense on requirements prioritization and the conformance of programs and budgets to priorities established both in strategic plans and those identified by the combatant commands. The JROC provides final validation of capability requirements, review and approval of joint prioritization, and final adjudication of any other issues. The JROC is the process owner for the Joint

Capabilities Integration and Development System (JCIDS) and uses the process to fulfill its advisory responsibilities to the chairman in identifying, assessing, validating, and prioritizing joint military capability requirements. In the army, we were taught that if one is in charge, then take charge and fix what is broken. General Hyten did not fix what he described as a broken acquisition system as he departed after a long and highly distinguished military career.

During a recent Leaders' Summit on Climate, Secretary of Defense Lloyd J. Austin III stated, "Today, no nation can find lasting security without addressing the climate crisis. We face all kinds of threats in our line of work, but few of them truly deserve to be called existential. The climate crisis does." He also said, "Climate change is making the world more unsafe, and we need to act" (Vergun, 2021).

I would hope that the seniormost official in DOD is more focused on the national security issues facing the US—such as the rising capabilities and threats posed by Russia, China, Iran, and the DPRK—and how to defend against hypersonic missiles designed to thwart our very expensive integrated air and missile defense systems, each capable of carrying nuclear warheads speeding toward US-defended assets at speeds between Mach 5 and Mach 20, which is 3,836 miles per hour to 24,696 miles per hour. Another critical national security issue facing our nation is the need to modernize the nation's strategic nuclear arsenal, which cost estimates say will be $1 trillion (McCausland, 2021). These senior leaders could publicly discuss the implications of Article 7 of China's National Intelligence Law passed in 2017, which requires organizations to include commercial companies to assist in Chinese espionage with direct implications to national security since it means that virtually every commercial endeavor could be collecting information and sharing it with Chinese intelligence. For example, this includes privately owned Chinese companies such as 23andMe DNA testing, which can be used to target biological weapons to specific groups and nationalities, to DJI UAS, which can be used to share video surveillance where they are being flown to cheap Huawei cell phones and towers popping up all over the globe, which can share cell conversations and data passed.

When one considers the immense number of critical infrastructure facilities in the US, estimated at over thirty thousand, it is obvious that the US cannot protect even a minority number of them. One option to counter the threat posed by UAS within the homeland would be the removal of the stated policies by federal agencies to withhold C-UAS systems from the civilian corporations that are the actual owner-operators of many of those facilities. The three-year security study by the Nuclear Regulatory Commission (NRC) determined that UAS were not a threat to US nuclear power plants, and it further recommended that the owner-operators of the ninety-six commercial US nuclear power plants should not be allowed to procure sensors or C-UAS systems to defend themselves. This study sent a message, unfortunately, it was the wrong message. The 1990 Iraq War may very well be the last time that the US will face a stupid enemy. Saddam Hussein allowed the US and its coalition of allied nations to build up its military force in theater. Future adversaries will undoubtedly have far greater resources and may be far more dangerous, both in equipment capabilities, the capacity, willpower, and determination to use their military forces to conduct a first strike to stifle US military actions. Many of our present-day adversaries paid close attention to how the US invaded Iraq, its tactics, strategies, and strengths but also its vulnerabilities and dependencies on low earth orbit satellites. Putin was a KGB officer stationed in East Germany when the collapse occurred. He referred to the collapse of the USSR as the "greatest geopolitical catastrophe of the century" (Taylor, 2018). Since 1990, Putin has focused Russian efforts on rebuilding a superpower capable of countering the US. Both Russia and China have poured considerable resources and expense into countering US strengths and attacking its vulnerabilities. They have expended significant resources in their pursuit of AI technologies to offset the US military advantages and to gain economic and military superiority. They are doing this to take advantage of what is an increasingly much-faster-paced and significantly more complex modern battlefield. Russia and China are developing AI to improve the independence, interoperability, and survivability of unmanned systems. They see AI-enhanced systems as replacing human soldiers during the next two decades. AI is an obvious way to reduce the lengthy and slow human oversight with autonomous AI that can make thousands of decision possibilities, possibly even millions, in the same amount of time as a human would make one. AI can incorporate what is referred to as the Joint All-Domain Command and Control (JADC2), a project that DOD is currently developing, which will incorporate information from multiple sensors while choosing from various kill chains to select the best kinetic or cyber platforms to launch an attack from against a single threat or multiple ones simultaneously. China has reported that it has combined its AI efforts with its hypersonic weapon systems; thus, as they test hypersonic missiles during routine testing, AI will learn how to predict and defeat inbound hypersonic missiles should the US use them against Chinese military targets.

Iran, the DPRK, Russia, and China have all been working on weapon and cyber systems to counter US strengths, both in the cyber realm, on the land, in the air, undersea, and in space. A few of these game-changing technologies have been recently unveiled in a demonstration of their technological superiority and can-do mental acuity over the US with the Chinese fractional orbital ballistic system, a hypersonic maneuvering glide vehicle, and Russian Tsirkon hypersonic missiles and Zircon hypersonic cruise missiles. Just as our competitors may be sharing their hypersonic technologies—Russia with India, China with the DPRK and, to a lesser extent, Iran with Houthi rebels—it is entirely conceivable that they would, in fact, share UAS technologies or end items of interest such as AI software with their proxies as long as they used it against the US or its interests and allies. In the end, it is the focus of one political system, the will of its leaders, and the depth of one's desire to finish on top vying for dominance against another. Even if we discount the efforts to share technological systems with another country or proxy, another area of concern for protecting critical infrastructure must be that of the inadvertent UAS operator. Take for example a grandfather who is demonstrating the capabilities of the new UAS to his grandson, and they are flying near a nuclear power plant. What a nice bonding experience to show little Johnny how well the UAS flies and how well the camera system works as they zoom over startled workers and security personnel, all wondering what is going on. Through carelessness, a hobbyist could lose control and fly into a transformer, power line, or another electronic box that could conceivably short out the flow of electricity, resulting in a loss of power for a neighborhood or business area. Currently, federal law or regulations may indeed prohibit critical infrastructure facilities from purchasing, training, and authorizing their security personnel to operate C-UAS against perceived threats. There are five federal entities with FAA C-UAS use exceptions: DHS, CBP, US Coast Guard (USCG), US Secret Service (USSS), and Federal Protective Service (FPS).

UAS poses a significant threat to all echelons of governments, corporations, and the public. UAS have proven their worth to the DOD in combat for many decades; they are now well-established with governments, corporations, and individual citizens. Their ability to provide clandestine surveillance, espionage, and armed attacks is so well established that there is no retreating on the use of this platform; in fact, the technologies and strategies for use have migrated to UUS and USS. FBI director Chris Wray testified before a Senate committee in 2017 that "drones are an increasing threat in the United States. Two years ago, this was not a problem. A year ago, it was an emerging problem. Now it is a real problem. So, we are quickly trying to up our game" (Dunkey, 2021). While our government falters on C-UAS strategies and the authorization to procure and distribute C-UAS equipment to multiple sectors of US critical infrastructure, they have yet to begin to digest the threat posed by USS, on land and atop the water in our rivers and lakes, and UUS to critical infrastructure. The DOT is focusing its efforts on figuring out how to utilize unmanned systems on roadways that were engineered for manned vehicles and not autonomous systems. The threat posed by unmanned cars, trucks, and semitractor trailers has yet to be addressed. The USCG has yet to develop the same for the threat posed by USS traveling atop the water, traversing the twenty-five thousand miles of inland waterways in the country. The navy is researching UUS operations and the opportunities for offensive and defensive operations are truly immense. Two decades ago, military members used to say that for the price of a fighter plane, a nation could buy fifty or more cruise missiles; thus, they had become a cheap air force. Today, the analogy with UUS is stark. For those small nations who simply cannot afford a large capital fighting ship, they can purchase one hundred or more autonomous UUS that can be deployed off of smaller ships, even fishing vessels, seek out an enemy's fleet, and engage them much as torpedoes do today; or they could conduct surveillance, reconnaissance of an enemies' ports, or act as mines in any harbor or the open seas. They can be launched ahead of submarines to search out enemy submarines and clear a path for the manned submarine. They are relatively cheap, disposable, and unmanned. They are rapidly on track to become a cheap replacement naval force or a naval augmentation force akin to the aircraft carriers in World War II.

We are facing the highest number of cyberattacks witnessed in history. Our critical infrastructure is at its highest degree of risk than ever before due in large part to the interconnected nature of our digital world. If one were to pay attention to the successes of cyberattacks, it is apparent that bad actors and strategic competitors would certainly target sectors of US critical infrastructure in the event of an attack upon or disruption to the homeland while pursuing regional conflicts around the globe, such as Ukraine for Russia and Taiwan for China. According to a 2020 Verizon DBI Report, "70 percent of the breaches in 2019 were perpetrated by external actors; of those, 64 percent were financially motivated, and 5 percent were espionage" (Moy, 2020). Cybercrime has no overriding sense of morality; rather, it is a revenue generator.

> "According to the FBI, since the COVID-19 life-altering pandemic began, cybercrime has quadrupled and the World Health Organization has reported a five-fold increase in cyberattacks directed against its staff and email scams targeting the public at large." (Moy, 2020)

The meteoric rise in Internet use, information technology, and AI has resulted in a corresponding rise in cybercriminal activities. The original inventors of the Internet and subsequent investors in information technologies that allow for work and social networking could not have envisioned the threat posed by cybercriminals and nation-states that now permeate the Internet. We must pool resources and increase the pursuit of transformational strategies to better secure the digital realm. The US critical infrastructure represents the front line of the cybersecurity entity. We must draw a red line in the sand on this issue and focus interagency and international efforts and resources on it. Failure to accomplish this will mean that much like Ukraine has suffered nearly a decade of cyberattacks from Russian hackers, we will lose our primacy in the world and society will be laid bare to our opponents.

There have always been and always will be threats to the homeland, but for the first time since the two world wars, we have internal strife that is as threatening or more than that posed by our enemies. During Brian Williams's politically charged farewell address as he signed off from MSNBC after two decades as a spokesman, he stated that he was leery of "the darkness on the edge of town that's spread to the main roads and highways and neighborhoods, the local bar, the bowling alley, the school board, and the grocery store" and further demanding that they "be answered for." Finally he shamed voters for having sent to Congress, people who've "decided to join the mob…to burn it all down with us inside. That should scare you to no end as much as it scares an aging volunteer fireman" (Houck, 2021). In 1858, Lincoln was speaking of the ills of slavery when he stated, "A nation divided against itself cannot stand" (Lincoln, 1858). According to Victor Davis Hanson, a senior fellow at the Hoover Institution:

> "America is declining as a society into a danger zone, and one of the main culprits is a refusal to prosecute crime and equally execute the nation's laws. The woke ideology is a very evil ideology because it's cruel. It's mean-spirited. And we haven't talked about that. But if that's what it is and it won't end until the people start identifying it like that… It's cruelty because it has a history

> throughout the centuries, and it doesn't end well. Academics...call this "systems collapse," where all of a sudden, a successful society suddenly doesn't follow its tradition and rules and things happen that people cannot believe, such as empty shelves or you're not able to buy meat or you go to fill up your car, and it's a hundred dollars...or people getting shot in the street.... And all of these things start to unwind the society. So, the point is that it can't continue. We fear the government and so the government's lost credibility, morality, and we've got to restore it by all of these DA's [restarting to] prosecute crimes. You have to be arrested. You have to be indicted if you're guilty. You have to serve your time to regain confidence and make it safe to be a human again."

Hanson also said, "When you look at the career of a Fauci, a Milley,...a Comey, a McCabe.... These aren't the people that we trust" (Grossman, 2022).

Another self-inflicted problem for the national security of the US is that of congressional leadership and responsibilities. According to an unnamed senior defense official:

> "The failure to pass a defense appropriations bill for fiscal year 2022 means that DoD will be restricted in how it operates for the year, and it will also stifle new initiatives to deal with the array of growing threats to the U.S." (Deptula, 2022)

Deptula shared his opinion which was substantiated by damning facts. He stated:

> "The only logical explanation is that there is a failure by the Congress, and by extension the

American people, to understand the gravity of the threats facing the U.S., which are growing stronger every day. It is truly shocking that Russia's preparations to launch the largest invasion in Europe since World War II and increasingly aggressive actions by China have done nothing to motivate Congress to pass this essential legislation. After years of counterinsurgency and counterterrorism operations in Afghanistan, Iraq, and elsewhere in the wake of 9/11, the Department of Defense finally woke up to the realization that it had neglected appropriate focus on the more significant threats of China and Russia. As just one example, the U.S. Air Force operates a geriatric force that is becoming more so every day. Its bombers and tankers are 60 years old; trainers over 50; fighters and helicopters over 40. For comparison, the average U.S. commercial airliner is about 10 years old—and they don't pull 9 times the force of gravity daily, as do our fighters. The Air Force Air Superiority Flight Plan says about the path we are on. The Air Force's projected force structure in 2030 is not capable of fighting and winning against the array of potential adversary capabilities. Unfortunately, the threat of a year-long continuing resolution puts the Defense Department's reconstitution effort on pause for a year, undermining our nation's defense as a result. Secretary of Defense Lloyd Austin summed up the impact by saying that a year-long continuing resolution, would erode the U.S. military advantage relative to China, impede our ability to innovate and modernize, degrade readiness, and hurt our people and their families. And it would offer comfort to our enemies, disquiet to our allies, and unnecessary stress to our work-

> force. The first responsibility of our government is the security of the American people. It's right there in the preamble of our Constitution, that the federal government was established to "provide for the common defense," (Deptula, 2022)

The fiscal year 2022 began without a budget rather than a continuing resolution that is set to expire without congressional consensus on a way forward on appropriations. Thus, the DOD is preparing for the possibility of a full year of operations under a continuing resolution. During a recent hearing by the "House Appropriations Committee's Defense Subcommittee, appropriators rightly acknowledged that a full-year continuing resolution would make our military less agile and curtail our ability to prepare for current security challenges" (Wynne, 2022). An abrogation of their responsibilities "will create a domino effect that will harm U.S. national security for years to come by damaging the growing unmanned systems industry" to address this new era of strategic competition vis-a-vis unmanned systems (Wynne, 2022). An example of the detrimental effect of non-action by Congress, the navy's new Unmanned Campaign Plan was developed to support the National Defense Strategy to build upon the emergence of unmanned system capabilities, yet past developments and investments envisioned to grow upon previous successes will fall behind our strategic competitors it continues to operate at the fiscal year 2021 funding levels, which were one-third of what was budgeted for the fiscal year 2022. These cuts represent significant losses of "time, capital, and investments that the defense-industrial base made in technology, supply base, workforce, supply chain and infrastructure based on the DoD's vision for the future" (Wynne, 2022).

There is a well-known reference to a nation's "death from a thousand cuts," and it is in reference to the demise of Athens to what was the world's second strongest nation, Sparta, as the sole superpower of ancient times. According to General Hyten, if we are afraid of failure, we must also be afraid then of acknowledging vulnerabilities and engaging in a rapid endeavor to develop counter systems to close or e due to our actions to stop Nazi Germany and Japanese imperialism in World War II. To remain a superpower, we must place the national security of our nation and its citizens at the very forefront of our nation's priorities.

Abbreviations

ACH	automated clearing houses
AMRAAM	advanced medium-range air-to-air missile
ARPANET	Advanced Research Projects Agency Network
ASW	anti-submarine warfare
ATM	automated teller machines
BTU	British thermal unit
C-UAS	counter-unmanned aerial system
CBP	Customs and Border Patrol
CBRN	chemical, biological, radiological, and nuclear
CIA	Central Intelligence Agency
CISA	Cyber Infrastructure Security Agency
COTS	commercial off-the-shelf systems
CWMD	countering weapons of mass destruction
DARPA	Defense Advanced Research Projects Agency
DHS	Department of Homeland Security
DDOS	distributed denial of service
DJI	Da-Jiang Innovations
DNS	domain name server
DOC	Department of Commerce
DOD	Department of Defense
DOI	Department of Interior
DOJ	Department of Justice
DoS	denial of service
DOT	Department of Transportation

DPRK Democratic People's Republic of Korea
EF education facilities
EMP electromagnetic pulse
EMS emergency medical services
EPA Environmental Protection Agency
EPIA Egg Products Inspection Act
FAA Federal Aviation Administration
FBI Federal Bureau of Investigation
FDA Federal Drug Administration
FFDCA Federal Food, Drug, and Cosmetic Act
FMIA Federal Meat Inspection Act
FOBS fractional orbital bombardment system
FOIA Freedom of Information Act
FSIS Food Safety and Inspection Service
FSLTT federal, state, local, territorial, tribal
FPS Federal Protective Service
GEO geosynchronous orbit
GMD Ground-based Midcourse Defense
GPS global positioning system
GSA General Services Administration
HE MD High Energy Laser Mobile Demonstrator
HEU highly enriched uranium
HGV hypersonic glide vehicle
HHS Department of Health and Human Services
IAMD integrated air and missile defense
ICBM intercontinental ballistic missile
IoT Internet of things
IP Internet protocol
ISIS Islamic State
ISR intelligence, surveillance, and reconnaissance
IT information technology

JADC2	Joint All-Domain Command and Control
JWC	joint warfighting concept
LEO	low Earth orbit
MCM	mine countermeasures
MEO	medium Earth orbit
MERS	Middle East respiratory syndrome
MIT	Massachusetts Institute of Technology
MitM	man in the middle
MW	megawatt
NASAMS	national advanced surface-to-air missile
NCR	National Capital Region
NIPP	National Infrastructure Protection Plan
NMI	National Monuments and Icons
NORAD	North American Aerospace Defense Command
NRC	Nuclear Regulatory Commission
NSA	National Security Agency
NSSE	National Special Security Event
OSD	Office of the Secretary of Defense
PLC	programmable logic controllers
PPD	Presidential Policy Directive
PPIA	Poultry Products Inspection Act
R&D	research and development
RADO	regimental air defense officer
SARS	severe acute respiratory syndrome
SCADA	supervisory control and data acquisition
SHORAD	short-range air defense
SLTT	state, local, tribal, and territorial
SSA	Sector-specific agency
SQL	server query language
TCP	Transmission Control Protocol
TSA	Transportation Security Administration

THAAD	Terminal High Altitude Area Defense
TW	terawatt
UAS	unmanned aerial system
USCG	US Coast Guard
USDA	US Department of Agriculture
USDT	US Department of the Treasury
USS	unmanned surface system
USSS	US Secret Service
UUS	unmanned undersea system
VCJCS	Vice Chairman of the Joint Chiefs of Staff
WMD	weapon(s) of mass destruction

References

Accenture Strategy. n.d.. "How the US Wireless Industry Powers the US Economy." https:// www.accenture.com/_acnmedia/pdf-74/accenture-strategy-wireless-industry-powers-us-economy-2018-pov.pdf.

Admin, B. (2021). "10 High Profile Cyber Attacks in 2021." https://cybermagazine.com/top10/10-high-profile-cyber-attacks-2021.

American Chemistry Council. (2011a). "Snapshot of the US Chemistry Industry in the United States." Washington, DC: American Chemistry Council.

American Chemistry Council. (2011b). "Chemistry Industry Facts and Figures." Washington, DC: American Chemistry Council.

American Chemistry Council. (2011c). "Energy. American Chemistry Council." Arlington, VA: American Chemistry Council.

American Chemistry Council. (2009). "Guide to the Business of Chemistry 2009." Arlington, VA: American Chemistry Council.

American Gaming Association. (2015). "Groundbreaking New Research Reveals Impressive Magnitude of US Casino Gaming Industry." http://www.gettoknowgaming.org/news/ groundbreaking-new-research-reveals-impressive-magnitude-us-casino-gaming-industry.

American Hotel & Lodging Association. (2014). 2014 Lodging Industry Profile. http://www. ahla.com/content.aspx?id=36332.

Americans for the Arts. n.d.. "Arts & Economic Prosperity 5." https://www.americansforthearts. org/sites/default/files/aep5/PDF_Files/ARTS_Brochure_Mockup.pdf.

Andrews, E. (2019). "Who Invented the Internet?" https://www.history.com/news/who-invented-the-internet.

Associated Press. (2021). "Sweden's spy agency probes drones over 3 nuclear plants." https:// abcnews.go.com/International/wireStory/sweden-puzzled-drones-spotted-nuclear-power-plants-82304055.

American Society of Civil Engineers. (2017). "2017 Infrastructure Report." https://infrastructurereportcard.org/tag/2017-infrastructure-report-card/.

Baksh, M. (2022). "FBI: Ransomware Attackers Have Code to Halt Critical Infrastructure." https://www. nextgov.com/cybersecurity/2022/02/fbi-ransomware-attackers-have-code-halt-critical-infrastructure/361808/.

Ball, M. (2022). "AI Software for Autonomous Litter Collection Vessels." https://www. unmannedsystemstechnology.com/2022/01/ai-software-for-autonomous-litter-collection-vessels/?utm_content=buffer6214f&utm_medium=social&utm_source=linkedin.com&utm_campaign=buffer.

Barrett, B. (2021). "A Drone Tried to Disrupt the Power Grid. It Won't Be The Last." https://www. wired.com/story/drone-attack-power-substation-threat/.

Bhutada, G. (2022). "Visualizing Nuclear Power Production by Country." https://elements.visual capitalist.com/visualizing-nuclear-power-production-by-country/.

Boutros, D. A., LCDR. (2015). *Operational Protection from Unmanned Aerial Systems: Drones, UAS Proliferation, ISR Platforms, UAS Capabilities and Vulnerabilities, Current Protective Measures, Protection Shortfalls.* UAS Navy War College, Newport, RI: Progressive Management Publications.

Bowman, B., J. Thompson, and R. Brobst. (2021). "China's surprising drone sales in the Middle East." https://www.defensenews.com/opinion/2021/04/23/chinas-surprising-drone-sales-in-the-middle-east/.

Boyer, S. (2010). *SCADA: Supervisory Control and Data Acquisition.* USA: International Society of Automation, 179. ISBN 978-1-936007-09-7.

Building Owners and Managers Association (BOMA) International. (2012). "Where America Goes to Work: The Contribution of

Office Building Operations to the Economy." http:// www.boma.org/industry-issues/state-local-issues/Documents/2011_BOMA_Econ_Impct _FINAL%20Proof%20for%20 print.pdf.

Bumgarner, J. (2010). "Computers as Weapons of War." *IO Journal.* https://web.archive.org/web/ 20111219174833/http:/www.crows.org/images/stories/pdf/IOI/IO%20Journal_Vol2Iss2_0210.pdf.

Bureau of Labor Statistics. (2014). United States Department of Labor. Current Employment Statistics. http://www.bls.gov/web/empsit/ceseeb1a.htm.

Carroll, C. (2011). "Cone of silence surrounds US cyberwarfare." *Stars and Stripes.* https:// www.stripes.com/news/cone-of-silence-surrounds-u-s-cyberwarfare-1.158090.

Cleveland, C. and C. Morris. (2013). *Handbook of Energy: Chronologies, Top Ten Lists, and Word Clouds.* Elsevier Science, 44. ISBN 978-0-12-417019-3.

Clines, F. (1991). "End of the Soviet Union; Gorbachev, Last Soviet Leader, Resigns; US Recognizes Republics' Independence." https://www.nytimes.com/1991/12/26/world/end-soviet-union-gorbachev-last-soviet-union-leader-resigns-us-recognizes-republics.html.

Cluley, G. (2022). "Nine-year-old kids are launching DDoS attacks against schools." https://www.bitdefender.com/blog/hotforsecurity/nine-year-old-kids-are-launching-ddos-attacks-against-schools/.

Collins, C. and D. Franz. (2018). "The Long Shadow of A. Q. Khan: How One Scientist Helped the World Go Nuclear." https://www.foreignaffairs.com/articles/north-korea/2018-01-31/long-shadow-aq-khan.

Combs, C. and M. Slann. (2009). *Encyclopedia of Terrorism, Revised Edition.* New York: Infobase Publishing.

Congressional Budget Office. (2018). "Monthly Budget Review: Summary for Fiscal Year 2018." https://www. cbo.gov/publication/54647

CTIA. n.d.a. "Annual Wireless Industry Survey." https://www.ctia.org/industry-data/ctia-annual-wirelessindustry-survey.

CTIA. n.d.b. "Wireless Snapshot." https://www.ctia.org/docs/default-source/default-documentlibrary/ctia-wireless-snapshot.pd.

CTIA. (2017). "Ericsson Mobility Report 2017." https://www.ericsson.com/assets/local /mobilityreport/documents/2017/ericsson-mobility-report-june-2017.pdf.

Cybersecurity Infrastructure Security Agency. n.d.. "Dam Sector." https//www.cisa.gov/dams-sector

Cybersecurity Infrastructure Security Agency. (2010). "Defense Industrial Base Sector-Specific Plan: An Annex to the National Infrastructure Protection Plan." https://www. cisa.gov/sites /default/files/publications/nipp-ssp-defense-industrial-base-2010-508.pdf.

Cybersecurity Infrastructure Security Agency. (2013). "National Infrastructure Protection Plan (NIPP) 2013: Partnering for Critical Infrastructure Security and Resilience." https://www.cisa.gov/publication/nipp-2013-partnering-critical-infrastructure-security-and-resilience.

Cyber Infrastructure Security Agency. (2015a). "Commercial Facilities Sector Specific Plan." cisa.gov/sites/default/files/publications/nipp-ssp-commercial-facilities-2015-508.pdf.

Cyber Infrastructure Security Agency. (2015b). "Communications Sector-Specific Plan: An Annex to the NIPP 2013." https://www.cisa.gov/sites/default/files/publications/nipp-ssp-communications-2015-508.pdf.

Cyber Infrastructure Security Agency. (2015c). "Critical Manufacturing Sector-Specific Plan." https://www.cisa.gov/sites/default/files/publications/nipp-ssp-critical-manufacturing-2015-508.pdf.

Cyber Infrastructure Security Agency. (2015d). "Dams Sector-Specific Plan." https.www.cisa.gov /sites/default/files /publications/nipp-ssp-dams-2015-508.pdf.

Cyber Infrastructure Security Agency. (2015e). "Emergency Services Sector-Specific Plan: An Annex to the NIPP 2013." https://www.cisa.gov/sites/default/files/publications/emergency-services-sector-specific-plan-112015-508_0.pdf.

Cyber Infrastructure Security Agency. (2015f). "Energy Sector-Specific Plan." https://www.
cisa.gov/sites/default/files/publications/nipp-ssp-energy-2015-508.pdf.

Cyber Infrastructure Security Agency. (2015g). "Financial Services Sector-Specific Plan 2015." cisa.gov/sites/default/files/publications/nipp-ssp-financial-services-2015-508.pdf.

Cyber Infrastructure Security Agency. (2016). "Information Sector-Specific Plan; An Annex to the NIPP 2013." cisa.gov/sites/default/files/publications/nipp-ssp-information-technology-2016-508.pdf.

Cyber Infrastructure Security Agency. (2019a). "Chemical Sector Landscape." https://www.cisa. gov/sites/default/files/publications/Chemical%20Sector%20Landscape_compliant.pdf.

Cyber Infrastructure Security Agency. (2019b). "Chemical Sector Profile." www.cisa.gov/sites/ default/files/publications/Chemical-Sector-Profile_Final%20508.pdf.

Cyber Infrastructure Security Agency. (2019c). https://wwww.cisa.gov/sites/default/files /publications/nipp-ssp-critical-manufacturing-2015-508.pdf.

Department of Energy. (2010). "Renewable Energy Sources: A Consumer's Guide." https://www. eia.gov/energyexplained/renewable-sources/#tab2.

Department of Homeland Security. n.d.. "Science and Technology: Cybersecurity Programs." www.dhs.gov/science-and-technology/cybersecurity-programs.

Department of Homeland Security. (2020). "Protecting against the Threat of Unmanned Aircraft Systems (UAS): An Interagency Security Committee Best Practice, November 2020 edition." https:///www.Critical%20Infra%20Structure/Protecting%20Against%20the%20Threat%20of%20Unmanned%20Aircraft%20Systems%20November%202020_508c.pdf.

Deptula, D. (2022). "The Biggest Threat To US National Security Today Is the US Congress." https://www.forbes.com/sites/davedeptula/2022/02/10/the-biggest-threat-to-us-national-security-today-is-the-us-congress/?sh=2053e8113856.

Dorn, T. (2020). *Unmanned Systems: Savior or Threat.* Pennsylvania: Page Publishing.

Durando, J. (2015). "Traditions: Macy's Thanksgiving Day parade explained." https://usatoday. com/story/news/nation-now/2014/11/24/thanksgiving-traditions-macys-parade/19364279/

Drees, J. (2021). "Ransomware attack forces Indiana hospital to divert patients." https://www. beckershospitalreview.com/cybersecurity/ransomware-attack-forces-indiana-hospital-to-divert-patients.html.

Edmondson, C. (2021). "Senate Passes $768 Billion Defense Bill, Sending It to Biden." https:// www.nytimes.com/2021/12/15/us/politics/defense-spending-bill.html.

Ehrlich, R. (2013). *Renewable Energy: A First Course.* CRC Press, 219. ISBN 978-1-4665-9944-4.

Einhorn, B. and T. Shields. (2021). "Drones Take Center Stage in US-China War on Data Harvesting." https://news.bloomberglaw.com/tech-and-telecom-law/drones-take-center-stage-in-u-s-china-war-on-data-harvesting.

Environmental Protection Agency. (2016). "Sector Programs: Chemical Manufacturing." https:// archive.epa.gov/sectors/web/html/chemical.html

Federal Aviation Administration. n.d.. https://www.faa.gov/uas/getting_started/remote _id/industry/.

Fink, E. (2014). "This drone can steal what's on your phone." https://money.cnn.com/2014 /03/20/technology/security/drone-phone/.

Forrester Research, Inc. (2015). "US Online Retail Sales to Reach $370 Billion by 2017." https:// www.forrester.com/US+Onlin e+Retail+Sales+To+Reach+370+Billion+By+2017/-/E-PRE4764.

Freedberg, S. (2018). "Army Anti-Aircraft Stryker Can Kill Tanks Too." https://breakingdefense. com/2018/07/army-anti-aircraft-stryker-can-kill-tanks-too/.

Funaiole, M. and B. Hart. (2021). "Understanding China's 2021 Defense Budget." https://www.csis .org/analysis/understanding-chinas-2021-defense-budget.

Gatlin, S. (2022). "FBI: BlackByte ransomware breached US critical infrastructure." https:// breakingdefense.com/2022/02/is-the-goal-of-jadc2-to-connect-sensors-to-shooters-or-should-it-be-command-and-control/.

Gardner, T. (2016). "Eighth drone spotted in SRS skies." http:// www.postandcourier.com/ aikenstandard/news/eighth-drone-spotted-in-srs-skies/article_21cb0f5a-d958-5fcf-9586-ec22f70433ba.html.

Gilbert, D. (2014). "Cost of Developing Cyber Weapons Drops from $100M Stuxnet to $10K Ice Fog." https://www.ibtimes.co.uk/cost-developing-cyber-weapons-drops-100m-stuxnet-10k-icefrog-1435451.

Girard, B. (2021). "The Real Danger of China's National Intelligence Law." https://thediplomat. com/2019/02/the-real-danger-of-chinas-national-intelligence-law/.

Goure, D. (2021). "Neither Manned Nor Unmanned: The Future of Air Warfare Will Be About Teaming." https://www.realcleardefense.com/articles/2020/05/13/neither_manned_nor_unmanned_the_future_of_air_warfare_will_be_about_teaming_115283.html.

Grau, L. and C. Bartles. (2016). "The Russia Way of War." https://www.armyupress.army.mil/ Portals/7/Hot%20Spots/Documents/Russia/2017-07-The-Russian-Way-of-War-Grau-Bartles.pdf.

Gross, J. (2011). "A Declaration of Cyber-War." *Vanity Fair*. www.vanityfair.com/news/ 2011/03/stuxnet-201104.

Grossman, H. (2022). "Woke ideology is 'cruel,' 'evil,' and won't end well: Victor Davis Hanson."

https://www.foxnews.com/media/woke-ideology-cruel-evil-victor-davis-hanson.

Hambling, D. (2021). "Improvised Killer Drones, Russian Mercenaries And False Flags." https://www.forbes.com/sites/davidhambling/2021/12/29/improvised-killer-drones-russian-mercenaries-and-false-flags/amp/.

Hambling, D. (2020). "Dozens more mystery drone incursions over US nuclear power plants revealed." https://www.forbes.com/

sites/davidhambling/2020/09/07/dozens-more-drone-incursions-over-us-nuclear-power-plants-revealed/#84041d06296b.

Hamilton, D. (2009). "White Sands testing new laser weapon system." https://www.army.mil/ article/16279/white_sands_testing_new_laser_weapon_system.

Horton, A. (2021). "Top US military leader: 'I want to understand White rage. And I'm White.'"

https://www.washingtonpost.com/powerpost/republicans-joint-chiefs-chairman-critical-racetheory-congress/2021/06/23/84654c34-d451-11eb-9f29-e9e6c9e843c6_story.html.

Hunter, J. (2020). "How F-16 Testers Are Evolving the Jet's New Radar Beyond the Homeland Defense Mission?" https://www.thedrive.com/the-war-zone/35678/how-f-16-testers-are-evolving-the-jets-new-radar-beyond-the-homeland-defense-mission.

International Association of Amusement Parks and Attractions. (2015). "Amusement Park and Attractions Industry Statistics." http://www.iaapa.org/resources/by-park-type/amusement-parks-and-attractions/industry-statistics.

Iscrupe, L. (2020). "What's the difference between Bluetooth and Wi-Fi?" https://www.allconnect. com/blog/difference-between-bluetooth-and-wifi.

Israeli Ministry of Defense. (2022). "Israel to build national laser-based air defense systems against drone and missiles." https://www.unmannedairspace.info/counter-uas-systems-and-policies/israel-to-build-national-laser-based-air-defence-system-against-drones-and-missiles/.

Judson, J. (2019). "It's official: US Army inks Iron Dome deal." https://www.defensenews.com/ digital-show-dailies/smd/2019/08/12/its-official-us-army-inks-iron-dome-deal/.

Kalvapalle, R. (2016). "Rio de Janeiro welcomed 1.17 million tourists in two weeks." https://www. marca.com/en/olympic-games/2016/08/24/57bda7a0468aeb3e158b4596.html.

Kaylin, S. and S. Westall. (2019). "Costly Saudi defenses prove no match for drones, cruise missiles." https://www.thedrive.com/

the-war-zone/29918/heres-all-the-new-info-you-need-to-know-in-the-aftermath-of-the-saudi-oil-facilities-attacks.

Keizer, G. (2010). "Is Stuxnet the 'best' malware ever?" https://www.infoworld.com/article 2626009/is-stuxnet-the--best--malware-ever-.html.

Kelley, M. (2012). "Obama Administration Admits Cyberattacks Against Iran Are Part of Joint US-Israeli Offensive." https://www.businessinsider.com/obama-cyberattacks-us-israeli-against-iran-2012-6.

Kerrigan, S. (2020). "Top 21 Dams in the World That Generate the Highest Amount of Electricity."

https://interestingengineering.com/top-21-dams-in-the-world-that-generate-the-highest-amount-of-electricity.

Kim, L. 2021. "Russian Cyberattacks Present Serious Threat to US." https://www.npr.org/ 2021/07/09/1014512241/russian-cyber-attacks-present-serious-threat-to-u-s.

Kofman, M. and R. Connolly. 2019. "Why Russian Military Expenditures is much higher than Commonly Understood (As in China)." https://warontherocks.com/2019/12/why-russian-military-expenditure-is-much-higher-than-commonly-understood-as-is-chinas/.

Kroft, Steve. 2012. "Stuxnet: Computer worm opens new era of warfare. 60 Minutes." CBS News. https://www.cbsnews.com/8301-18560_162-57390124/stuxnet-computer-worm-opens-new-era-of-warfare/.

Kwan, C. 2019. "Rio Tinto completes rollout of world-first autonomous iron ore rail operations." https://www.zdnet.com/article/rio-tinto-completes-rollout-of-first-heavy-haul-long-distance-autonomous-rail-operation-in-the-world/.

Lincoln, A. 1858. Speech of Hon. Abraham Lincoln before the Republican state convention, June 16, 1858. Sycamore, IL: O. P. Bassett.

Lister, T. 2019. "The billions Saudi Arabia spends on air defenses may be wasted in the age of drone warfare." https://www.cnn.com/2019/09/19/middleeast/saudi-air-defense-analysis-intl/index.html.

Lock, S. 2020. “Casino industry—statistics and facts.” https://www.statista.com/topics/1053 /casinos/#dossierKeyfigures.

Looking Glass. n.d. “Understanding Cyber Risk for the Defense Industrial Base.” http:// lookingglasscyber.com/blog/security-corner/understanding-cyber-risk-for-the-defense-industrial-base/.

Lopez, C. 2021. “DoD Aims to Bring Industrial Base Back to the US, Allies.” https://www. defense.gov/News/News-Stories/Article/Article/2474015/dod-aims-to-bring-industrial-base-back-to-us-allies/.

Losey, S. 2022. “CISA warns of potential Russian cyberattacks as invasion fears mount.” https://link.defensenews.com/click/26714125.99804/aHR0cHM6Ly93d3cuYzRpc3JuZXQuY29tL2N5YmVyLzIwMjIvMDIvMTQvY2lzYS13YXJucy1vZi1wb3RlbnRpYWwtcnVzc2lhbi1jeWJlcmF0dGFja3MtYXMtaW52YXNpb24tZmVhcnMtbW91bnQv/6127a2427c5f9829851e6049B54d4f84a.

Lukeheart, A. 2022. “2022 Cyber Attack Statistics, Data, and Trends.” https://parachutetechs.com/2022-cyber-attack-statistics-data-and-trends/.

Lyngaas, S. 2021. “Drone at Pennsylvania electric substation was first to ‘specifically target energy infrastructure,’ according to federal law enforcement bulletin.” https://www. cnn.com/2021/11/04/politics/drone-pennsylvania-electric-substation/index.html.

Majumdar, D. 2017. “Raytheon has a Genius Plan to Make the Stinger Missile a Drone-Killer.” https://nationalinterest.org/blog/the-buzz/raytheon-has-genius-plan-make-the-stinger-missile-drone-20995.

McAfee. 2017. “McAfee Labs Threat Report: April 2017.” https://www.mcafee.com/enterprise /en-us/assets/reports/rp-quarterly-threats-mar-2017.pdf.

McCausland, J. 2021. “General Milley, critical race theory and why GOP’s ‘woke’ military concerns miss the mark.” https://www.nbcnews.com/think/opinion/general-milley-critical-race-theory-why-gop-s-woke-military-ncna1272558.

McMillan, R. 2010. "Siemens: Stuxnet worm hit industrial systems." Computerworld. https:// www.computerworld.com/article/2515570/siemens--stuxnet-worm-hit-industrial-systems.html.

Mizokami, K. 2019. "This is the ATV-Mounted Jammer that took down an Iranian Drone." https://www.popularmechanics.com/military/weapons/a28471436/lmadis-iranian-drone/.

Mizokami, K. 2021. "Autonomous Drones Have Attacked Humans. This Is a Turning Point." https://www.popularmechanics.com/military/weapons/a36559508/drones-autonomously-attacked-humans-libya-united-nations-report/.

Montenegro, K. 2021. "The world's first laser-armed drone Eren impressed with its successful shots in the tests." https://www.-aa-com-tr.translate.goog/tr/bilim-teknoloji/dunyanin-ilk-lazer-silahli-dronu-eren-testlerdeki-basarili-atislariyla-goz-dolduruyor/2444041?_x_tr_ sl=tr&_x_tr_tl=en&_x_tr_hl=en-US.

Moy, R. 2020. "Protecting critical infrastructure and distributed organizations in an era of chronic cybersecurity risk." https://www.securitymagazine.com/articles/93663-protecting-critical-infrastructure-and-distributed-organizations-in-an-era-of-chronic-cybersecurity-risk.

National Multifamily Housing Council. n.d. "Quick Facts: Resident Demographics." http://www. nmhc.org/Content.aspx?id=4708#What_type_of_structure.

National Oceanic and Atmospheric Administration. n.d. "Coronal Mass Ejections." https://www. swpc.noaa.gov/phenomena/coronal-mass-ejections.

National Retail Federation. 2013. Annual Report. https://nrf.com/annualreport2013/.

National Strategy for Physical Protection of Critical Infrastructures and Key Assets. 2003. "The National Strategy for the Physical Protection of Critical Infrastructures and Key Assets." https://www.dhs.gov/xlibrary/assets/Physical_Strategy.pdf.

Navarro, J. 2021. "Box office revenue in the US and Canada 1980–2020." https://www.statista. com/statistics/187069/north-american-box-office-gross-revenue-since-1980/.

Newdick, T. 2022. "Bomblet Dropping Drones Are Now Being Used By Cartels In Mexico's Drug War." https://www.thedrive.com/the-war-zone/43847/bomblet-dropping-drones-are-now-being-used-by-cartels-in-mexicos-drug-war.

Newdick, T. 2021. "Russia's Predator-Like Drone Is Now Shooting Down Other Drones." https://www.thedrive.com/the-war-zone/43585/russias-predator-like-drone-is-now-shooting-down-other-drones.

Opall-Rome, B. 2017. "Raytheon: Arab operated Patriots Intercepted over 100 tactical ballistic missiles since 2015." https://www.defensenews.com/digital-show-dailies/dubai-air-show/ 2017/11/14/raytheon-saudi-based-patriots-intercepted-over-100-tbms-since-2015/.

Osborn, K. 2021. "The US Military is Working to Counter China's New Drone Mothership." https://nationalinterest.org/blog/reboot/us-military-working-counter-chinas-new-drone-mothership-196463.

Peck, M. 2021. "Turkey Reveals World's 1st Laser-Armed Drone (Don't Get Too Excited)." https://www.19fortyfive.com/2021/12/turkey-reveals-worlds-1st-laser-armed-drone-dont-get-too-excited/.

Perlroth, N. 2021. "This is How They Tell Me the World Ends." New York: Bloomsbury Publishing.

Price Waterhouse Cooper. 2018. "Global entertainment & media outlook 2017–2021: User experience takes centre stage." https://www.pwc.com/gx/en/industries/entertainment-media/outlook.html.

Plunkett Research, Limited. 2015. "Sports Industry, Teams, Leagues & Recreation Market Research." http://www.plunkettresearch.com/statistics/sports-industry/.

Pomerleau, M. 2022. "Russia and China devote more cyber forces to offensive operations than US, says new report." https://link.defensenews.com/click/26714125.99804/aHR0cHM6Ly93d3cuYzRpc3JuZXQuY29tL2N5YmVyLzIwMjIvM-DIvMTQvcnVzc2lhLWFuZC1jaGluYS1kZXZvdGUtb-W9yZS1jeWJlci1mb3JjZXMtdG8tb2ZmZW5zaXZlLW9w-

ZXJhdGlvbnMtdGhhbi11cy1zYXlzLW5ldy1yZXBvcnQv/6127a2427c5f9829851e6049Bc0513864.

Qiblawi, T. 2019. "Attacks have disrupted 5% of the world's oil production. Here's what you need to know." https://www.cnn.com/2019/09/16/middleeast/saudi-oil-attacks-explainer-intl/

Qiblawi, T. 2022. "A drone attack in Abu Dhabi could mark a dangerous turning point for the Middle East. Here's what to know." https://www.cnn.com/2022/01/18/middleeast/uae-abu-dhabi-houthi-yemen-explainer-intl/index.html.

Ranson, A. 2017. "The 2014 UAV threat to French nuclear power plants." *National Security & the Future*, 18(1/2): 125–142.

Reisner, Marc. 1986. "Cadillac desert: the American West and its disappearing water." New York, NY: Viking.

Reuters. 2019. "Explainer: How the Saudi attack affects global oil supply." https://www. reuters.com/article/us-saudi-aramco-oil-global-explainer/explainer-how-the-saudi-attack-affects-global-oil-supply-idUSKBN1W11WH.

Reuters. 2021. "Iraqi PM chairs security meeting after drone attack on residence." https://m-jpost-com.cdn.ampproject.org/c/s/m.jpost.com/breaking-news/explosion-heard-near-iraqi-pms-residence-in-baghdad-report-684231/amp.

Roblin, S. 2017. "The Coming Laser Wars." https://nationalinterest.org/blog/the-buzz/the-coming-laser-wars-21401.

Roblin, S. 2018. "The US Army Needs More Anti-Aircraft Weapons—and Fast." https:// warisboring.com/the-u-s-army-needs-more-anti-aircraft-weapons-and-fast/.

Roblin, S. 2019a. "Why US Patriot missiles failed to stop drones and cruise missiles attacking Saudi oil sites." https://www.nbcnews.com/think/opinion/trump-sending-troops-saudi-arabia-shows-short-range-air-defenses-ncna1057461.

Roblin, S. 2019b. "Chinese Drones Are Going to War All Over the Middle East and Africa." https://nationalinterest.org/blog/buzz/chinese-drones-are-going-war-all-over-middle-east-and-africa-74246.

Roblin, S. 2019c. "The Army's Big Six: America Plan to Wipe out Russia or China in a War." https://nationalinterest.org/blog/

buzz/armys-big-six-america-plan-wipe-out-russia-or-china-war-61862.

Rogoway, T. and J. Trevithick. 2020. "The night a mysterious drone swarm descended on Palo Verde Nuclear Power Plant." https://www.thedrive.com/the-war-zone/34800/the-night-a-drone-swarm-descended-on-palo-verde-nuclear-power-plant.

Rogoway, T. 2019. "Air Force's Secretive XQ-58A Valkyrie Experimental Combat Drone Emerges After First Flight." https://www.thedrive.com/the-war-zone/26825/air-forces-secretive-xq-58a-valkyrie-experimental-combat-drone-emerges-after-first-flight.

Rogoway, T. 2018. "This new video showing DARPA's master plan for its Gremlins Drones is awesome." https://www.thedrive.com/the-war-zone/20802/this-new-video-showing-darpas-master-plan-for-its-gremlins-drones-is-awesome.

Rogoway, T. 2017. "America's Starling Short Range Air Defense Gap and How to Close it Fast." www.thedrive.com/the-war-zone/13284/americas-gaping-short-range-air-defense-gap-and-why-it-has-to-be-closed-immediately.

Ryland, T. and S. Williamson. (2020). "What Was the US GDP Then?" https://www.measuring worth.com/datasets/usgdp/.

Sakharkar, A. 2021. "DARPA awards Phase 2 contracts for Manta Ray underwater drone." https://www.inceptivemind.com/darpa-awards-phase-2-contracts-manta-ray-underwater-drone/22587/.

Schaffer, A. 2021. "The Cybersecurity 202: Russian hackers haven't backed off, administration official acknowledges." www.washingtonpost.com/politics/2021/10/06/russian-hackers-havent-backed-off-administration-official-acknowledges/.

Silver, C. 2021. "The Top 25 Economies in the World." https://www.investopedia.com/insights /worlds-top-economies/.

Sir, R. 2021. "Do You Know About Vampire tap? Computer Anudeshak 2021." www.youtube. com/watch?v=WlDZxEbkU9Q.

Skillings, J. 2013. "Army laser weapon KOs mortar rounds." https://www.cnet.com/news/army-laser-weapon-kos-mortar-rounds/.

Southwell, H. 2021. "Everything from Ford Rangers to Giant Dump Trucks get automated in the Australian Mining Industry."

https://www.thedrive.com/news/39310/everything-from-ford-rangers-to-giant-dump-trucks-get-automated-in-the-australian-mining-industry.

Stark, H. 2011. "Mossad's Miracle Weapon: Stuxnet Virus Opens New Era of Cyber War." Der Spiegel. www.spiegel.de/international/world/mossad-s-miracle-weapon-stuxnet-virus-opens-new-era-of-cyber-war-a-778912.html.

Statista. 2021. "Banking, Finance, and Insurance in the US 2021." https://www.statista.com/ study/15822/banking-finance-and-insurance-in-the-us/.

Stein, R. 2018. "Museums as Economic Engines." https://www.aam-us.org/2018/01/19/ museums-as-economic-engines/.

Stein, V. 2021. "What is the speed of light?" https://www.space.com/15830-light-speed.html.

Symantec. 2018. "ISTR 24: Symantec's Annual Threat Report Reveals More Ambitious and Destructive Attacks." https://symantec-enterprise-blogs.security.com/blogs/threat-intelligence/istr-24-cyber-security-threat-landscape.

Taylor, A. 2018. "Putin Says He Wishes the Soviet Union Had Not Collapsed. Many Russians Agree." https://www.ndtv.com/world-news/russian-president-vladimir-putin-says-he-wishes-the-soviet-union-had-not-collapsed-many-russians-agr-1819331.

Tepperman, J. 2021. "The Most Serious Security Risk Facing the United States." https:// nytimes.com/2021/02/09/books/review/this-is-how-they-tell-me-the-world-ends-nicole-perlroth.html.

The Real Estate Roundtable. 2009. "Continuing the Effort to Restore Liquidity in Commercial Real Estate Markets." http://www.rer.org/uploadedFiles/RER/Policy_Issues/Credit_Crisis/2009_09_Restoring_Liquidity_in_CRE.pdf?n=8270.

The White House. 2013. "Presidential Policy Directive—Critical Infrastructure Security and Resilience." https://obamawhitehouse.archives.gov/the-press-office/2013/02/12/presidential-policy-directive-critical-infrastructure-security-and-resil.

The White House. 2003. "Homeland Security Presidential Directive (HSPD-7)." http://www. whitehouse.gov/news/releases/2003/12/20031217-5.html.

Tingley, B. 2022. "Autonomous Resupply Gliders Made Successful Deliveries On Their First Overseas Deployment." https://www.thedrive.com/the-war-zone/44111/autonomous-resupply-gliders-made-successful-deliveries-on-their-first-overseas-deployment.

"Transcript: Securing Cyberspace with Jen Easterly." 2021. *Washington Post Live* https://www. washingtonpost.com/washington-post-live/2021/10/05/transcript-securing-cyberspace-with-jen-easterly/.

Trevithick, J. 2021a. "Gremlins Drones could be rearmed inside their mothership transport aircraft." https://www.thedrive.com/the-war-zone/41042/gremlins-drones-could-be-reloaded-inside-their-mothership-transport-aircraft.

Trevithick, J. 2021b. "Surface To Air Missiles Now Needed to Protect Critical US Infrastructure During A Crisis." https://www.thedrive.com/the-war-zone/42788/critical-u-s-infrastructure-now-needs-surface-to-air-missile-protection-during-a-crisis.

Trevithick, J. 2019. "Tests For DARPA's Gremlins Drones Are All Laid Out but May Be Headed to New Venue." https://www.thedrive.com/the-war-zone/30260/tests-for-darpas-gremlins-drones-are-all-laid-out-but-may-be-headed-to-new-venue.

Tubbs, S. 2018. "Programmable Logic Controller (PLC) Tutorial, Siemens Simatic S7-1200." Publicis MCD Werbeagentur GmbH; 3rd ed.

Turton, W. and K. Mehrotra. 2021. "Hackers Breached Colonial Pipeline Using Compromised Password." https://www.bloomberg.com/news/articles/2021-06-04/hackers-breached-colonial-pipeline-using-compromised-password.

United Nations Security Council. 2021. "Final Report of the Panel of Experts on Libya established pursuant to Security Council resolution 1973 (2011)." https://undocs. org/S/2021/229.

US Bureau of Economic Analysis. 2020. "Outdoor Recreation Satellite Account, US, and States, 2020." https://www.bea.gov/news/2021/outdoor-recreation-satellite-account-us-and-states-2020.

US Census Bureau. 2015a. "Statistics of US Businesses." http://www.census.gov/econ/susb/.

US Census Bureau. 2015b. "Economic Census: Industry Snapshots: Primary Metal Manufacturing (NAIC 331)." http://thedataweb.rm.census.gov/TheDataWeb_HotReport2 /econsnapshot/2012/snapshot.hrml?NAICS=331.

US Census Bureau. 2015c. "Economic Census: Industry Snapshots: Primary Metal Manufacturing (NAIC 331)." http://thedataweb.rm.census.gov/TheDataWeb_HotReport2/ econsnapshot/2012/snapshot.hrml?NAICS=331.

US Census Bureau. 2015d. "Economic Census: Industry Snapshots: Electrical Equipment, Appliance, and Component Manufacturing (NAICS 335)." http://thedataweb.rm.census. gov/TheDataWeb_HotReport2/econsnapshot/2012/snapshot.hrml?NAICS=335.

US Census Bureau. 2015e. "Economic Census: Industry Snapshots: Transportation Equipment Manufacturing (NAICS 336)." http://thedataweb.rm.census.gov/TheDataWeb_HotReport2 / econsnapshot/2012/snapshot.hrml?NAICS=336.

US Congress. 2012. "America is under cyberattack: Why urgent action is needed." Hearing before the Subcommittee on Oversight, Investigations, and Management of the Committee on Homeland Security House of Representatives, One Hundred Twelfth Congress, Second Session, Serial No. 112-85. https://www.govinfo.gov/content /pkg/CHRG-112hhrg 77380/html/CHRG-112hhrg77380.htm.

US Department of Commerce. 2021. "National Institute of Standards and Technology: Manufacturing Industry Standards." https://www.nist.gov/el/applied-economics-office/manufacturing/manufacturing-industry-statistics.

US Department of Labor. 2013. "May 2013 National Industry-Specific Occupational Employment and Wage Estimates." http://www.bls.gov/oes/current/naics4_711200.htm.

US Energy Information Administration. 2021. "What is US electricity generation by energy source?" https://www. eia.gov/tools/faqs/faq.php?id=427&t=3.

US Energy Information Administration. 2020. "Nuclear Explained: US Nuclear Energy." https://www.eia.gov/energyexplained/nuclear/us-nuclear-industry.php

US Nuclear Regulatory Commission. 2019. "Backgrounder on Nuclear Waste." https://www. nrc.gov/reading-rm/doc-collections/fact-sheets/radwaste.html.

US Travel Association. 2014. "Travel Exports: Driving Economic Growth and Creating American Jobs." https://www.ustravel.org/sites/default/files/page/2009/09/2014_ Export_Report-PDF-FINAL.pdf.

Vavra, S. 2021. "North Korean Hackers Caught Snooping on China's Cyber Squad." https:// www.thedailybeast.com/north-korean-hackers-caught-snooping-on-chinas-cyber-squad.

Venable, J. and L. Ries. 2021. "DJI Placed on the Entity List for Human Rights Abuses but Concerns About Data Security Should Not Be Overlooked." https://www.heritage.org/ cybersecurity/commentary/dji-placed-the-entity-list-human-rights-abuses-concerns-about-data.

Venditti, B. 2021. "Visualizing China's Evolving Energy Mix." https://elements.visual capitalist.com/visualizing-chinas-energy-transition/.

Vergun, D. 2021. "Defense Secretary Calls Climate Change an Existential Threat." https://www. defense.gov/News/News-Stories/Article/Article/2582051/defense-secretary-calls-climate-change-an-existential-threat/.

Watson, A. 2021. "Global book publishing revenue 2018–2023." https://www.statista.com/ statistics/307299/global-book-publishing-revenue/.

"Weird but wired." 2007. *The Economist*. https://www.economist.com/asia/2007/02/01/weird-but-wired.

Wei, W. 2018. "Casino Gets Hacked Through Its Internet-Connected Fish Tank Thermometer." https://thehackernews.com/2018/04/iot-hacking-thermometer.html.

Weisgerber, M. 2017. "Stinger Missiles Can Now Shoot Down Small Drones." https://www. defenseone.com/technology/2017/06/stinger-missiles-can-now-shoot-down-small-drones/138322/.

Westcott, K. 2017. "Media and Entertainment industry trends: US." https://www2.deloitte.com/ us/en/pages/technology-media-and-telecommunications/articles/media-and-entertainment-industry-outlook-trends.html.

World Nuclear Organization. 2021. "Nuclear Power in Egypt." https://world-nuclear.org/ information-library/country-profiles/countries-a-f/egypt.aspx.

World Population Review. 2021. "Defense Spending By Country 2021." https://worldpopulation review.com/country-rankings/defense-spending-by-country.

Wynne, B. 2022. "A yearlong continuing resolution will hinder unmanned systems integration." https://www.c4isrnet.com/opinion/commentary/2022/01/14/a-yearlong-continuing-resolution-will-hinder-unmanned-systems-integration/.

Xuanzun, L. 2021. "China conducts test flight for airborne unmanned swarm carrier." https:// www.globaltimes.cn/page/202104/1220474.shtml.

Yakowicz, W. 2021. "US Gambling Revenue to break $44 Billion Record in 2021." https://www.forbes.com/sites/willyakowicz/2021/08/10/us-gambling-revenue-to--break-44-billion-record-in-2021/?sh=71e0bc88677b.

Yang, Z. 2018. "China's laser weapons: future potential, future tensions?" https://www. rsis.edu.sg/rsis-publication/rsis/co18093-chinas-laser-weapons-future-potential-future-tensions/#.YaVA0cfMJ3g.

Zengerle, P. 2021. "Biden Administration Proceeding With $23 Billion Weapon Sales to UAE." https://www.usnews.com/news/world/articles/2021-04-13/exclusive-biden-administration-proceeding-with-23-billion-weapon-sales-to-uae.

Zwijnwenburg, W. 2014. "Drone-tocracy? Mapping the Proliferation of Unmanned Systems." https://paxforpeace.nl/news/blogs/drone-tocracy-mapping-the-proliferation-of-unmanned-systems.

About the Author

Dr. Terence M. Dorn was born in Ankara, Turkey, and as the son of an air force chief master sergeant, he traveled worldwide and eventually attended college in Nebraska, where he earned two baccalaureate of art degrees in business administration and sociology from Bellevue College. Dr. Dorn joined the army in 1985, excelled in the army's air defense branch, and was deployed seven times to various combat zones. Two noteworthy assignments included serving as a speechwriter to the chief of staff of the army and military assistant to the secretary of defense. In 2013, he was inducted into the Army Officer Candidate School Hall of Fame and retired in 2014.

While serving the nation as an army officer, Dr. Dorn earned a master of arts degree in international relations from Boston University and a master of science degree in national security strategy from the National War College. In 2020, he completed his PhD in business administration from Northcentral University. His area of specialization was in Homeland Security, and his dissertation was titled A Phenomenological Study Examining the Vulnerabilities of U.S. Nuclear Power Plants to Attack by Unmanned Aerial Systems. Dr. Dorn authored Unmanned Systems: Threat or Savior and Quotes: The Famous and Not So Famous in honor of his father. He is currently writing The Future is Unmanned. In addition, Dr. Dorn wrote Countering the Threat Posed by Unmanned Systems Armed with WMDs while working at the Department of Homeland Security and an implementation plan to turn that strategy into reality. Dr. Dorn's civilian honors include induction into the Academy of Criminal Justice Sciences, Alpha Phi Sigma, Iota Pi; the National Society of Leadership and Success, Sigma Alpha Pi; and the International Business Honor Society, Delta Mu Delta.

CPSIA information can be obtained
at www.ICGtesting.com
Printed in the USA
LVHW011656050723
751517LV00019B/444/J